The role of additives in plastics

The role of additives in plastics

L. MASCIA
Lecturer in plastics technology,
University of Aston in Birmingham

A HALSTED PRESS BOOK

JOHN WILEY & SONS · NEW YORK

© L. Mascia 1974

First published 1974
by Edward Arnold (Publishers) Ltd.
London

Published in the U.S.A.
by Halsted Press, a Division
of John Wiley & Sons, Inc.
New York

ISBN: 0 470-57410-0
Library of Congress Catalog Card No. 73-14098

Printed in Great Britain by Fletcher & Sons Ltd
London & Norwich

PREFACE

The desire to compile this book was initiated partly by the author's experience in the teaching of plastics technology and partly as a result of a survey carried out among industrial technologists. A major difficulty experienced by students appeared to be associated with the lack of a comprehensive textbook on additives relating industrial practice to the underlying scientific principles. This is understandable to some extent since first, in technology many principles are not capable of accurate scientific explanation, (a situation which is also reflected in many sections of this book) and second, to fully understand the behaviour of plastics it often requires crossing the barrier of some quite distinct disciplines. A technologist can be guided however, by simple interdisciplinary models and mechanisms in his search for solutions to practical problems. This theme forms the basis on which this book was written, making maximum use of general principles which are applicable to industrial problems. Although the emphasis is on the benefits accrued from additives, some of the more important aspects of adverse interactions have been included.

This book is not intended primarily for specialists but for those wishing to acquire a basic knowledge of the use of additives in plastics materials, hence excessive literature review has been avoided and, to preserve the identity of a monograph, the information has been presented in a descriptive manner wherever possible.

The principles of mechanical property modifications have been emphasized to a greater extent than others since this is a major area in which chemists and engineers will have to share responsibility most for the successful advancement of plastics technology as a discipline comparable to metallurgy. The level of treatment in general is such that it should be well within the grasp of both workers in industry engaged in development of polymeric compositions and students in technical colleges, polytechnics and universities, who study plastics technology as part of their general curriculum.

Finally, the subdivision of chapters has been made with the view that a reader may wish to concentrate on a particular group of additives without having to make frequent cross-references.

The author wishes to acknowledge the help received from Dr. M. Tahan and Dr. J. T. Barnby, in the gathering of some of the information contained in Chapters 3 and 6 and to Mr. J. E. Proctor and Mr. S. Ludlow for reading the scripts and making useful comments.

L. Mascia

CONTENTS

1 GENERAL ASPECTS OF ADDITIVES FOR PLASTICS

Introduction

Since the very early stages of the development of the polymer industry it was realized that useful products could only be obtained if certain additives were incorporated into the polymer matrix. This process is normally known as 'compounding'. The term compounding was first used in the rubber industry and introduced by Goodyear in 1839 when he discovered that the addition of sulphur to a raw rubber stock gave much improved products.

About thirty years later a similar situation was experienced with the very first plastics product: cellulose nitrate. The use of large quantities of solvents was found to be necessary to shape the raw material, while the addition of camphor considerably improved the toughness of the finished products.

1.1 Definition and classification of additives

The term 'additives' is used here to describe those materials which are physically dispersed in a polymer matrix without affecting significantly the molecular structure of the polymer. Cross-linking agents, catalysts, etc., normally used in thermosetting systems are therefore excluded.

Additives used in plastics materials are normally classified according to their specific function, rather than on a chemical basis. It is also convenient to classify them into groups and to subdivide them further

1

according to their more precise function:

(i) Additives which assist processing
- (a) Processing stabilizers
- (b) Lubricants — Internal / External
- (c) Processing aids and flow promoters
- (d) Thixotropic agents

(ii) Additives which modify the bulk mechanical properties
- (a) Plasticizers or flexibilizers
- (b) Reinforcing fillers
- (c) Toughening agents

(iii) Additives used to reduce formulation costs
- (a) Particulate fillers
- (b) Diluents and extenders

(iv) Surface properties modifiers
- (a) Antistatic agents
- (b) Slip additives
- (c) Anti-wear additives
- (d) Anti-block additives
- (e) Adhesion promoters

(v) Optical properties modifiers
- (a) Pigments and dyes
- (b) Nucleating agents

(vi) Anti-ageing additives
- (a) Anti-oxidants
- (b) U.V. stabilizers
- (c) Fungicides

(vii) Others
- (a) Blowing agents
- (b) Flame retardants

1.2 Technological requirements of additives

(a) *Compatibility and mobility of additives*

The most important requirement of any additive is that it should be effective, for the purpose for which it has been designed, at an economic level.

Improvements in one property can, however, lead to deteriorations in others and, consequently it is the overall performance of an additive in a given formulation which determines the final choice.

Furthermore, the effectiveness of compounding additives depends also on the correct procedure of incorporation into the polymer matrix. The appropriate physical form of the additive in the polymer matrix depends in turn on the mechanism by which it exerts its function.

Complete compatibility (i.e. mutual miscibility at molecular level) and mobility or diffusibility of the additive molecules within the polymer matrix are essential if the action of the additive is such that any or all the molecules of the system are to interact with each other.

Total incompatibility and immobility of the additive molecules are desirable when the additive exerts its function at a supermolecular level, i.e. the action of the additive is derived from its intrinsic physical properties in the bulk or macroscopic form.

Partial compatibility is required when a strong affinity between polymer and additive is to be exerted at the interface. This is best achieved when at the interface the physical properties change smoothly from those of the polymer to those of the additive, even if such a change is to take place over small distances, e.g. a few molecular layers.

The compatibility and diffusibility of additives in polymeric compositions is normally assessed by 'trial and error' methods; a practice which is still likely to be used for some time in the future. The principal reason for this situation is to be attributed to the lack of sound scientific alternatives. The theories put forward so far, in fact, have been primarily developed for solvent/polymer systems and may have limited applicability to the case of additive/polymer mixtures, where the polymer is normally the major constituent.

In addition to this, the number of additives available and their possible combinations in polymeric systems are enormous and their compositions are continually changing. However the basic principles of solution thermodynamics can be used to provide a rough guide for the assessment of the relative compatibility of additives and polymers. It is intended to give here only a brief account of this and consequently oversimplifications will be made.

Entropy of mixing   This can be obtained by constructing a lattice model consisting of additive and polymer molecules of equal size and shape so that their position in the lattice is interchangeable. Using the Boltzmann equation, we can obtain an approximate expression for the entropy in terms of the respective volumetric fractions of the two components,

$$\Delta S_M = k \, ln \, \Omega \tag{1.1}$$

where k is the Boltzmann constant, and Ω is the total number of possible arrangements for the additive and polymer molecules.

If, n_0 = total number of molecules

$\quad n_1$ = number of polymer molecules

$\quad n_2$ = number of additive molecules

then

$$\Omega = \frac{n_0!}{n_1! \cdot n_2!} \tag{1.2}$$

and, therefore

$$\Delta S^M = k \, ln \, \frac{n_1!}{n_1! \cdot n_2!} = k \, (ln \, n_0! - ln \, n_1! - ln \, n_2!) \tag{1.3}$$

Using the Sterling approximation, $ln \, n! = n \, ln \, n - n$, and substituting in the above equation we obtain

$$\Delta S^M = k[n_0 \, ln \, n_0 - n_1 \, ln \, n_1 - n_2 \, ln \, n_2] \tag{1.4}$$

In terms of molar volumetric fractions, where

$\quad \varphi_1 = n_1/n_0$ molar volumetric fraction of the polymer

$\quad \varphi_2 = n_2/n_0$ molar volumetric fraction of the additive

the above equation becomes:

$$\Delta S^M = k[n_0 \, ln \, n_0 - n_1 \, ln \, n_0 - n_1 \, ln \, \varphi_1 - n_2 \, ln \, n_0 - n_2 \, ln \, \varphi_2]$$

and since $(n_0 - n_1 - n_2)ln \, n_0$ is zero, then

$$\Delta S^M = -k[n_1 \, ln \, \varphi_1 + n_2 \, ln \, \varphi_2] \tag{1.5}$$

Heat of mixing   The heat of mixing can be obtained by considering the changes in internal energy (ΔE_{12}), which take place when the additive and polymer are mixed together, relative to their respective internal energy in their pure state.

According to Hildebrand and Scott[1] this energy change can be obtained from the expression

$$\Delta E_{12}^M = (n_1 v_1 + n_2 v_2) \left[\left(\frac{E_1^v}{v_1} \right)^{\frac{1}{2}} - \left(\frac{\Delta E_2^v}{v_2} \right)^{\frac{1}{2}} \right]^2 \varphi_1 \varphi_2 \tag{1.6}$$

where v_1 is the volume of one polymer molecule, v_2 is the volume of one additive molecule, the quantities $\Delta E^v/v$ are the respective energy of vaporization per unit volume, and the values of $(\Delta E/v)^{1/2}$ are called 'solubility parameters' and denoted by the symbol δ.

If V_1 is the volume occupied by the polymer and V_2 is the volume occupied by the additive, equation (1.6) can be re-written as

$$\Delta E_{12}^M = \varphi_1 \varphi_2 \, V_1 \, (\delta_1 - \delta_2)^2 \qquad (1.7)$$

since $V_1 \gg V_2$.

If the mixing process is considered to take place without changes in volume, $\Delta V^M = 0$, and since $\Delta H^M = \Delta E^M + P\Delta V^M$, then equation (1.3) also represents the heat of mixing, i.e. $\Delta H^M = \Delta E^M$.

Free energy of mixing   This quantity can now readily be computed from the entropy and heat of mixing equations,

$$\Delta F^M = \Delta H^M - T\Delta S^M$$
$$= \varphi_1 \varphi_2 \, V_1 (\delta_1 - \delta_2)^2 - kT(n_1 \, ln \, \varphi_1 + n_2 \, ln \, \varphi_2) \qquad (1.8)$$

According to the second law of thermodynamics, mixing (like any other spontaneous process) occurs if there is a decrease in free energy, i.e. ΔF^M must be negative or ΔF^M positive.

In the solid state when the concentration of additive is small we can only expect a small increase in entropy. It follows, therefore, that these restrictions will only allow a relatively small increase in ΔH. Hence δ_1 and δ_2 will have to be as close as possible. Somewhat larger differences in δ values are allowable when there is possibility of H-bonding between the two components, since this would result in a decrease in ΔH.

We have now reached the conclusion that it is possible to obtain a rough indication of whether an additive is miscible in the polymer considered from a knowledge of the relative values of the heat of mixing or from a knowledge of their solubility parameters δ_1 and δ_2. Solubility parameters for a variety of polymers and additives, plasticizers in particular, are now becoming available in the literature and handbooks, and it may be useful to consult these when selecting additives for plastics compositions.

(b) *Migration and consumption of additives*

The additive must neither volatilize out of the matrix during processing nor exude to the surface during service. This implies that the additive must have a low vapour pressure at high temperatures and must not aggregate, i.e. precipitate or crystallize out of the polymer matrix, on ageing, and leave behind a fine film of additive deposits. This phenomenon is called 'chalking'. The additive must not be extractable by liquids with which the host polymeric composition may come into contact during finishing operations, neither must it exude out during its life in service. The two latter phenomena are known as 'bleeding' and 'blooming' respectively. Not only would these phenomena produce aesthetically objectionable effects and contaminate liquids and other products in contact with the plastics component, but the loss of additives from the system would inevitably reduce its efficiency.

Bleeding and blooming phenomena are obviously related to the kinetics of diffusion and consequently are dependent on parameters such as compatibility of the additive with the polymer, molecular size of the additive, physico-chemical interactions between additive and polymer molecules, configuration of polymer chains and intermolecular voids etc.

Hence insoluble inorganic additives, such as pigments, fillers etc. are unlikely to bleed or bloom, whereas soluble low-molecular-weight plasticizers are more likely to exude to the surface during processing and subsequent ageing and may even constitute a vehicle for the migration of other soluble additives such as processing stabilizers.

When the additive in question functions by interacting with either the polymer, another additive or environmental agents, it would be desirable if it could be made auto-regeneratable so that its effectiveness would not depend on its previous history. There are indications that, to some extent, an auto-regeneration mechanism operates with some 'synergistic' stabilizer systems (see later).

(c) *Health hazards of additives*

An additive must not have any damaging effects on the health of the personnel engaged in compounding and processing operations nor on that of the consumer, especially when the plastics component is intended for packaging of food products or for toys. Legislation is particularly stringent with respect to protection of consumers against toxic effects derived from extracted additives.

The British Plastics Federation has laid down certain recommendations for the plastics industry, which may be summarized as follows[3]:

(1) If an ingredient of a plastics material cannot be extracted by foodstuff with which it is in contact, it does not constitute a toxic hazard.

(2) If a material is found in a food as a result of its contact with plastics it may constitute a toxic hazard if in itself it is toxic in the biological sense.

(3) Acute toxic levels are unlikely to be realized in practice. It is possible, however, that injurious effects may be produced by repeated small doses of material extracted from plastics and therefore it is the accumulative effect which should be used in assessing the hazard.

(4) The toxic hazard is a function of the chronic toxicity and of its extractability from a plastics material under service conditions.

(5) Extractability tests must be carried out with the foodstuffs themselves or a range of representative extractants under the most severe conditions likely to be incurred in practice. The results must then be combined with the data on chronic toxicity expressed by their 'toxicity-factor' and a given 'toxicity quotient', which is a measure of the hazard.

(6) As an additional safety factor all 'Schedule I' poisons should never be in contact with foodstuffs.

The results of tests are expressed by

$$Q = \frac{E \times 1000}{T}$$

where (1) Q = toxicity quotient; (2) T = toxicity factor of the extracted material, which is a figure assessed by the Federation on the examination of available data and approximates to the maximum daily oral dose expressed in mg/kg body weight which can be tolerated for 90 days by groups of animals without producing any detectable toxic effect; (3) E = weight in grams of extracted material per specific volume or surface area of the original sample depending on whether this is a thick or thin section of material.

A plastics material is considered satisfactory for use in contact with foodstuffs if the sum of all 'toxicity quotients' of extracted ingredients is less than 10.

The extractants usually used are:

Distilled water, 5% Na_2CO_3, 5% acetic acid, 50% ethyl alcohol, olive oil + 2% oleic acid or paraffin oil, 5% citric acid.

Extraction tests are carried out under three conditions:

(a) 45 °C for 24 hours — intermittent contact
(b) 60 °C for 10 days — prolonged contact
(c) 80 °C for 2 hours — intermittent contact with food

1.3 Unavoidable side-effects of additives: deterioration of dielectric properties

It is often the case that a combination of additives are incorporated into plastics to produce from a few basic polymers a wide range of 'grades', in order to meet all the service requirements which a product may demand.

It is not surprising, therefore, that chemical and physical interactions which would lead to undesirable side-effects may take place among the various additives. Some of these interactions will be dealt with in later chapters by reference to specific examples. Furthermore there are certain polymer properties which are invariably impaired as a result of additive interactions; notably among these are the dielectric properties. This point will be illustrated by considering the effects on dielectric losses and electric break-down voltage.

1.3.1 Dielectric losses

The disturbance of electric charges within a material and the resultant surface charges when the material is subjected to an external electric field is described in terms of dielectric constant ϵ and polarization P.

The polarization of a material is defined as the change in charge density on the plates of a condenser when the material is used as the dielectric in place of vacuum, i.e. $P = q_{(vac)} - q_{(diel)}$. Within the material the displacement of charges will neutralize one another, and therefore polarization occurs only at the two surfaces in contact with the plates of the condenser. The dielectric constant, on the other hand is defined as the relative increase in capacitance (C) of the condenser or the relative decrease in voltage gradient (V) when the charge density is kept constant, i.e.

$$\epsilon = \frac{C(\text{diel})}{C(\text{vac})} = \frac{V(\text{vac})}{V(\text{diel})} \qquad (1.9)$$

Thus as a result of introducing the material between the plates of the condenser instead of vacuum, the field strength is reduced inversely proportionally to its dielectric constant, i.e. $E_{(vac)} = \epsilon \cdot E_{(diel)}$. The total polarization P is the algebraic sum of several components known respectively as 'interfacial' or 'space-charge polarization' P_i, 'dipolar' or 'orientation polarization' P_d, 'atomic polarization', P_a, and 'electronic polarization' P_e.

Interfacial polarization arises from macroscopic movements of ions within the material. Orientation polarization arises from the distortion and rotation of permanent dipoles, which in the case of polymers consist of asymmetric pendant side groups or portions of the molecular backbone chains.

Atomic polarization arises from very small movements of atoms within a molecular structure, and finally, electronic polarization is associated with movements of electrons.

All polarization processes are time dependant owing to constraints on the various displacements imposed by neighbouring constituents of the structure of the material. This time dependance can be described by exponential equations of the type $P = P_0(1 - e^{-t/\tau})$ where τ is a characteristic 'relaxation time' for each particular charge displacement. Relaxation times are very small for atomic and electronic polarization ($\cong 10^{-13}$ and $\cong 10^{-15}$ s)[4], but are considerably larger for dipolar and interfacial polarizations ($\cong 10^{-8}$ and 10^{-2} s respectively). In alternating electric fields, polarization also varies periodically with time but the internal constraints to rapid movements of charges causes a dissipation of energy, normally in the form of heat. So whereas in vacuum the electric current would form a phase angle of $90°$ with the voltage, polarization of the dielectric will decrease *pro-rata* the phase angle, as shown in Fig. 1.1. Consequently the effective electric current I has an imaginary component I_l, in phase with the voltage, known as the loss current, and a real component I_c which corresponds to the current intensity in absence of any losses (e.g. in vacuum) and it therefore forms a $90°$ phase angle with the voltage.

From Fig. 1.1 we obtain $\tan \delta = I_l/I_c$, which is called the 'loss tangent'.

Accordingly we can represent the dielectric constant in complex form $\epsilon^* = \epsilon' - j\epsilon''$, where ϵ'' is called the 'loss factor', and it can be shown that $\epsilon''/\epsilon' = \tan \delta$.

The power losses which occur in dielectrics are proportional to ϵ''

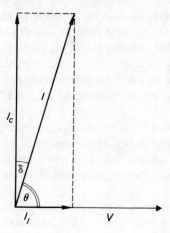

Fig. 1.1 Vector diagram showing the phase angle θ and the loss angle δ (I_l is in phase with V).

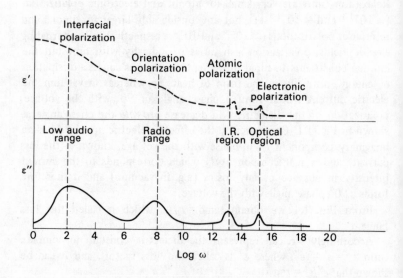

Fig. 1.2 Relative contributions to overall polarization and dielectric losses[6].

and tan δ, from which the terms loss factor and loss tangent are derived. Both ϵ'' and ϵ' are frequency dependant and vary in magnitude according to the type of polarization which takes place as shown in Fig. 1.2. In electrical applications one is normally concerned with frequencies less than 10^{10} Hz and therefore there will be no contribution from both atomic and electronic polarizations to dielectric losses. Hence it is the orientation and interfacial polarization which must be minimized to obtain high quality insulating materials

Consequently most polymers used in plastics, in their 'pure' state and at relatively low temperatures, are not likely to give rise to appreciable losses in the low audio frequency range and with non-polar polymers, such as polyolefines, there will be relatively low losses over the whole range of frequencies of practical importance

1.3.2 Dielectric strength

The dielectric strength of an insulating material is expressed in terms of minimum voltage which causes a permanent loss of dielectric properties (by converting it into a conductive material) under specified conditions. This results from a breakdown in the chemical structure of the material mainly as a result of thermal and environmental degradation processes.

The heat generated as a result of dielectric losses can raise the temperature of polymers (owing to their low thermal diffusivity) to levels at which there will be sufficient thermal energy to cause molecular cleavage (see p. 25). This can lead to the formation of double bonds and ultimately to conjugated macromolecular structures.

Both temperature rise and chemical structure breakdown will increase the conductivity of the material which will eventually lose completely its insulating characteristics.

Plastics based on phenolic materials e.g. PF's, polycarbonates etc., can produce[7] readily conductive paths by forming highly conjugated graphitic structures

Environmental agents, air, moisture or other ionic species present either on the surface or in the bulk of the polymer can accelerate the rate of temperature rise and structure breakdown and, therefore, will reduce the dielectric strength of the material.

Since the temperature will rise at different rates and the chemical breakdown process is obviously also rate dependant, it is to be expected that the dielectric strength of a plastics material is not constant. Neither would it be surprising to find the a.c. breakdown voltages to be appreciably lower than the d.c. voltage values measured under similar conditions.

1.3.3 Adverse effects of additives

It was shown in the preceeding sections that it is the polarity of the constituent groupings of polymer molecules and the presence of mobile ionic species which are mostly responsible for the dielectric losses of materials and, which at the same time, will contribute to dielectric breakdown.

Additives may increase losses either because of their intrinsic ionic and/or polar nature or because they may absorb water, which increases further the dipolar and ionic constituents of the system. Their adverse effect on dielectric breakdown may result from the higher losses which will increase the rate of temperature rise, and if they are hygroscopic they may accelerate the decomposition of polycondensate systems, e.g. heterochain linear polymers and thermosets.

Incompitable additives, especially those with a high surface free-energy (e.g. inorganic fillers and pigments) may increase the interfacial polarization by adsorption and immobilization of ionic and polar species which will in turn decrease the effective internal charge neutralization. Furthermore they may give rise to the formation of irregular microvoids at the interface, which may trap harmful gaseous matter such as oxygen, water, etc. At the same time on the asperities of these microcracks there will be a high charge density which constitute points of weakness regarding the initiation of the breakdown process. Oxygen, in fact, ionizes and forms ozone under these conditions, which causes a very rapid thermal oxidative breakdown. Although it is generally believed that non-hygroscopic and contamination-free fillers such as mica flakes, etc., may improve the dielectric strength of phenolic compounds by interrupting the continuity of the conductive path through the material, this is only likely to be true when very high filler loadings are used and the compound is cured under very high pressures. Otherwise the interfacial discharge breakdown will predominate and the dielectric strength will therefore be reduced[5].

1.4 Methods of incorporation of additives into polymer matrices

The manner in which additives and polymers are blended together or 'compounded' is determined by the following factors:

(a) physical form and melting characteristics of the polymer considered,
(b) physical form and concentration of additive to be used,
(c) degree of dispersion or solubilization of the additive in the final mix,
(d) physical form of the plastics raw material or 'compound' to be produced.

It is understood that in those cases where more than one method of compounding is suitable for the given system considered, the ultimate choice may be influenced by other factors such as equipment availability, tradition, etc. The general criterion adopted for the mixing of polymeric systems and additives is shown in Table 1.1 and the

Table 1.1   Criterion for the mixing of additives with polymeric systems[3]

Operation	Criterion	Examples
Dissolution	High shear, circulation	Dissolving solids and gums in liquids
Solid suspensions	Circulation, low flow velocity	Slurries and pastes
High viscosity fluid blending	Circulation, low flow velocity	Lattices and adhesives
Paste and melt agitation	Kneading, internal mixing	Used with most thermoplastics
Solid blending	Tumbling, circulation	Dry powder blending and pre-mixing

mixing procedures normally used are given in schematic form in Table 1.2.

For the latter it can be seen that the complexity of the procedure adopted, and hence the economics, depends on the degree of dispersion required. It must be emphasized, however, that there is a tendency nowadays to increase the mixing efficiency of processing equipment so that more economical compounding procedures such as 'tumble mixing' can be adopted.

There are of course other possibilities which have not been considered here, especially those which may be used for liquid systems. Details of these are outside the scope of this book and can be found elsewhere[7]

Table 1.2 Normal mixing procedures for plastics compounding

Physical form of the polymer	Physical form of the additive	Pre-mixing operation	Further mixing operation	Physical form of the 'blend' or 'compound'	Type of dispersion	Notes:
Powder	Liquid or powder	High speed vortex blending	Kneading and extrusion	Granular product	Good	a) A 'Masterbatch' is a 'granular compound' containing a high concentration of additives.
	Liquid or powder	High speed vortex blending	—	'Dry' powder blend	Medium	
	Additive concentrate in polymer powder	Tumble mixing	—	'Tumbled' powder blend	Fair/poor	b) A 'powder concentrate' is a concentrated mixture of polymer and additive, both in the form of powder.
Granules or chips	Masterbatch	Internal melt mixing	Kneading and extrusion	Granular product	Excellent	
	Masterbatch	Tumbling	—	Granular mix	Fair	
	Powder	Tumbling	—	Dry 'coloured' granules	Fair/poor	Powder concentrates are normally obtained by high speed vortex blending.
	Paste concentrate	Roll-mixing or sigma blade blending	—	Paste	Excellent	
Liquid	Powder	Ball-mixing or roll-mixing	—	Paste	Good	

14

There is the additional case for the mixing of continuous or long fibres with liquid polymer systems to be considered. Normally this is done by impregnation of the fibres or by spraying the resin or polymer solution directly on to the fibres prior to moulding.

It should be noted that the 'degree of dispersion' is normally described qualitatively, e.g. good, poor, fine, etc. In the case of dispersion of solid additives (e.g. pigments) in liquid systems another term, 'dispersion stability', is used to describe, in a similar manner, the extent to which the original degree of dispersion is maintained in the product during handling or storage.

It is also important to appreciate that although the performance of an additive depends on its degree of dispersion in the polymer matrix, it is often difficult to assess this by optical methods, especially if two distinct phases are absent. Hence it is not surprising that the degree of dispersion is often implied, rather than measured directly, from the assessment of its relative effectiveness with respect to those properties of the polymer which the additive is intended to modify.

References

1. HILDEBRAND, J. H. and SCOTT, R. L., *The Solubility of Non-Electrolytes*, Reinhold, 1950.
2. GARDNER, R. J., *Plastics*, June, 1964, p. 74.
3. SIMONDS, H. R., *The Encyclopaedia of Plastics Equipment*, Reinhold, 1964, p. 13.
4. RITCHIE, P. D., *Physics of Plastics*, Plastics Monograph, Iliffe Books, 1965, p. 290.
5. *Ibid.*, p. 309.
6. BAIRD, M. E., *Electrical Properties of Polymeric Materials*, The Plastics Institute, 1973, p. 2.
7. *Encyclopaedia of Polymer Science and Technology*, Vol. 4, Interscience, 1965, p. 118.

2 ADDITIVES WHICH ASSIST PROCESSING

The factors which most of all determine product quality and output in plastics processing are the resistance of the polymer to thermal degradation, the frictional behaviour of the melt on the metal surface of the processing equipment and the melt viscosity of the polymer.

High thermal stability in polymers enables processing to be carried out at high temperatures and therefore increases the output, as a result of the reduction in melt viscosity. It also improves product quality as a result of the elimination of the deleterious effects of degradation on physical properties and of objectionable discolourations. A control of the frictional behaviour of the polymer melt on metal surfaces is essential as it determines the residence time of polymer molecules at the interface and hence their likelihood of undergoing chemical changes.

The term 'degradation' is used here to denote any chemical process which alters the chemical structure of the polymer in a manner which leads to a deterioration in its physical properties.

2.1 Processing stabilizers

The degradation of polymers brought about by the effects of heat and/or oxygen normally occurs by a free-radical mechanism:

(a) Initiation step: Production of free radicals

$$RH \text{ (Polymer)} \xrightarrow{\text{energy}} R* + H*$$

(b) Propagation step: Radicals interaction with polymer chains

$$R* + O_2 \longrightarrow ROO*$$
$$ROO* + RH \longrightarrow ROOH + R*$$

(c) Termination step: Deactivation of free radicals

$$R* + R* \longrightarrow R\text{--}R$$
$$R* + ROO* \longrightarrow ROOR$$
$$ROO* + ROO* \longrightarrow ROOR + O_2$$

The overall effect of these reactions is that oxygen combines with polymer chains to form carbonyl compounds which will accumulate and give rise to characteristic yellow and brown discolourations. Further decomposition of hydroperoxides can take place in the propagation step producing chain scission reactions,

Recombination of chain radicals in the termination step, on the other hand, can produce cross-links in the polymer structure.

Oxidation reactions may be accelerated by the presence of heavy metal ions, e.g. Co^{2+}, which are known to catalyse the decomposition of hydroperoxides. The overall thermal oxidative process depends on the ease of hydrogen abstraction from polymer chains, the order of which is:

CH in α positions to aromatic groups $\left(\text{HC--} \bigcirc \right)$ > Allyl–CH groups >

tertiary C–H groups > secondary H–CH groups > primary H_2CH groups

Breakdown of chemical bonds cannot be prevented by external means, i.e. by means of additives, and the inherent stability of polymers may only be improved by introducing suitable alterations in the chemical structure. Additives, on the other hand, can only exert an arrestive (or retardative) action on the degradation process. Arrestive stabilization can be achieved by any of the following mechanisms:

(a) intervention directly with the degradation reactions to produce inactive species and to reduce the concentration of reactive species, so that the resultant overall rate for the degradation process is decreased,

(b) removing, deactivating or promoting competition for sources that have a catalytic action on the degradation process.

Owing to the inevitable entrapment of air during processing, oxidation reactions always play a very important role in processing degradation. Even in the case of halogenated polymers, where the major process is the direct de-halogenation (see later), the presence of oxygen can accelerate the overall degradation reactions. Consequently most stabilizers have been developed around oxidation reactions.

Details of stabilization mechanisms are not given here; the interested reader will find extensive treatments in other publications and textbooks.

Processing stabilizer systems can be conveniently classified in the following manner:

2.1.1 Primary stabilizers: antioxidants

The action of antioxidants is rather complex and interpretations in the literature are often controversial. The general consensus of opinion is that antioxidants interrupt the chain reactions by combining with the propagating free-radicals species and thus forming non-reactive products. They act according to the mechanism

$$\sim\!\!\sim\!\!-P\cdot + AH \longrightarrow \sim\!\!\sim\!\!-PH + \overset{\cdot}{A} \text{ (inactive radical);}$$

where $\sim\!\!\sim\!\!-P\cdot$ is the propagating species and may include any of the following groupings:

$$\sim\!\!\sim\!\!-CH_2\cdot, \quad \sim\!\!\sim\!\!-\underset{R}{CH}\cdot, \quad \sim\!\!\sim\!\!-\underset{R}{C}\cdot, \quad \sim\!\!\sim\!\!-CH_2O\cdot, \quad \cdot OH, \quad \text{etc.}$$

Most commercial primary stabilizers are either hindered phenols or aromatic amines. A list of effective stabilizers is given in Table 2.1.

The lack of activity of the reaction products derives from their resonance stabilization and/or dimerization with other species. Thus with phenolics compounds:

(Stable products)

(Resonance stabilization)

and similarly with aromatic amines.

According to this mechanism the protection of the polymer against thermal oxidation ceases when all the antioxidant has been consumed by the chain propagation species. The period over which this protection takes place is called the 'induction time', (Fig. 2.1) and after this period the degradation reactions proceed as if no antioxidant had been added.

The efficiency of an antioxidant can be expressed in terms of increased induction time under conditions considered, or on the basis of its increased induction time over a wide range of temperatures, together with the activation energy for the process as derived from an Arrhenius plot (Fig. 2.2).

2.1.2 Secondary stabilizers: peroxide decomposers

A second way of protecting polymers from degradation is to remove peroxide radicals and to decompose hydroperoxides as they are formed.

Many sulphur and phosphorous compounds can in fact act in this way and the most common types are sulphides, thioethers, tertiary

Table 2.1 Typical antioxidants available commercially (Total levels of addition: approximately 0·2–0·75 p.h.p.)

Chemical name	Structure	Trade name and supplier	Comments
4,4'butylidene (6-tert-butyl m. cresol)		Santowhite Powder (Monsanto)	—
1,1,3-tris (2 methyl-4-hydroxy-5-tert-butylphenyl) butane		Topanol CA (I.C.I.)	Good ageing and processing stabilizer. Can be synergized with DLTD or DSTDP

Chemical name	Structure	Trade name	Comments
tetra tris methylene-3-(3,5'-di-ter-butyl-4'-hydroxy phenyl) propionate methane		Irganox 10 10 (CIBA-Geigy)	—
2,4,5-tris-(2-5 di-tibutyl-4-hydroxybenzyl)-1,3,5-trimethyl benzene		Ionox 330 (Shell)	Excellent process-ing stabilizer. Widely used in cables formulations. Dipoles perfectly balanced, hence little effect on dielectric losses.
Di(n-octadecyl)-3,5-di-t-butyl-4-hydroxy benzyl phosphate		Irganox 1093 (CIBA-Geigy)	Good processing stabilizer.

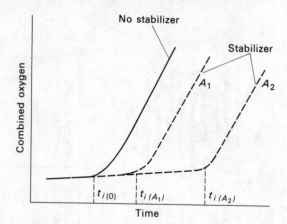

Fig. 2.1 Effect of stabilizers on the induction time of thermal degradation. Where $t_{i(0)}$ = induction time for unstabilized polymer: $t_{i(A_1)}$ and $t_{i(A_2)}$ = induction time for stabilized polymer; A_1 and A_2 denote either two different antioxidants, or the same antioxidant at two concentration levels.

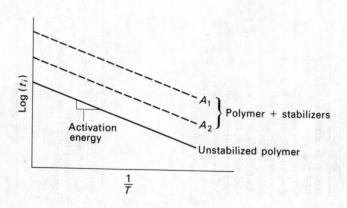

Fig. 2.2 Arrhanius plot of induction time for unstabilized and stabilized polymer.

phosphites and phosphonates (see Table 2.2). The mechanism of their reactions is complex but there seems to be some agreement that the peroxides are reduced to alcohols and are, therefore, deactivated in the manner shown below:

$$\sim\sim\text{CHOOH} + \text{P(OR)}_3 \longrightarrow \sim\sim\text{CHOH} + \text{O}=\text{P(OR)}_3$$

$$\sim\sim\text{CHOO}\cdot + \text{RSH} \longrightarrow \sim\sim\text{CHOOH} + \text{RS}\cdot$$

$$\sim\sim\text{CHOO}\cdot + \text{RS}\cdot \longrightarrow \sim\sim\text{CH}-\text{OOSR} \quad \text{(stable products)}$$

2.1.3 Chelating agents: metal deactivators

In the preceding pages it has been said that metals of variable valency can enhance degradation of polymers by accelerating the rate of decomposition of hydroperoxides to active radicals.

Complexing these metal ions with 'bulky' ligands creates steric hindrance around the metal ion and prevents its interaction with active sites of the polymer chains.

The presence of metal ion impurities in polymers can arise from many sources, e.g. catalyst residues of Ziegler-Natta or redox initiators, fillers, pigments, etc.

The most important class of chelating agents are compounds based on organic phosphines and phosphites, and higher nitrogenated organic compounds, e.g. melamine, bis-salicylidene diamines, oxamides, etc.

2.1.4 Special stabilizers for halogenated polymers

Halogenated polymers, such as those based on vinyl and vinylidene chloride, are very important commercially and present special problems with respect to degradation during processing. There are also auxiliary additives, e.g. flame retardants, impact modifiers, etc. which are based on halogenated products and may create further processing stabilization problems when incorporated into polymers. The area of stabilization of halogenated polymers has been exploited very widely over the past thirty years and as a result a very large number of stabilizers have been made available commercially.

Table 2.2 Typical secondary stabilizers available commercially
(Total level of addition: approximately 0·1—0·5 p.h.p.)

Chemical name	Structure	Trade name and supplier
Dilaurylthio-dipropionate	$S(CH_2CH_2-\overset{O}{\overset{\|}{C}}-O-C_{12}H_{25})_2$	Plastanox LTDP (American Cyanamid) and DLTDP (Robinson Bros.)
Distearylthio-dipropionate	$S(CH_2-CH_2-\overset{O}{\overset{\|}{C}}-OC_{18}H_{37})_2$	Plastanox STDP (American Cyanamid)
Tris(nonylphenyl) phosphite	$\left(C_9H_{19}-\bigcirc-O \right)_3 P$	Santowhite TNPP
Tris (mixed mono- and dinoylphenyl phosphite)	$\left(\right)_n$ structure with C_9H_{19} groups	Polygard (Uniroyal)

For a more comprehensive guide for the selection of stabilizers consult *Modern Plastics Encyclopaedia* (McGraw-Hill).

In addition to normal oxidative reactions described previously, a very rapid de-hydrohalogenation process can take place. With PVC (for instance) scission of C–Cl bonds occurs in the weakest points of the molecular chains, i.e. allylic or tertiary positions. The Cl* radicals will abstract hydrogen from adjacent CH groups and create another weak allylic C–Cl bond which becomes susceptible to scission, hence giving rise to HCl unzipping reactions:

$$-CH_2-CH-CH_2-CH-CH_2- \quad \xrightarrow{\text{energy}} \quad -CH_2-CH-CH_2-CH-CH_2-$$

with Cl below first and second CH of left structure; right structure with Cl below first CH, and + Cl*

$$-CH_2-CH-CH_2-CH=CH- \quad \xleftarrow[\text{by HCl}]{\text{(catalysed}} \quad -CH_2-CH-CH_2-CH=CH-$$

+ Cl* (left); Cl below, + HCl (right)

(weak allylic bond)

$$-CH_2-CH=CH-CH=CH-$$

+ HCl   etc. } conjugated unsaturation responsible for the early development of the polymer discolouration.

Therefore stabilizers for halogenated compounds must meet the following requirements:

(1) absorb and neutralize HCl evolved in order to arrest autocatalytic chain reactions and to prevent corrosion of processing equipment,
(2) prevent oxidation reactions and other free-radical processes,
(3) displace active, labile substituent groups, e.g. tertiary and allylic Cl atoms, with more stable substituents,
(4) disrupt conjugation in the residual polymer chains to prevent the formation of objectionable discolourations.

For many years stabilizer systems for polyvinyl chloride have been based on substances capable of neutralizing HCl, and these still constitute an important class of stabilizers. These stabilizer systems are based on basic lead salts and weak basic soaps (Table 2.1).

Stabilization by displacement of labile chlorine[1], on the other hand, has been accomplished and exploited in more recent years by means of heavy metal carboxylates, cadium and zinc soaps, and mercaptides (Table 2.3).

The mechanism for such substitution reactions is as follows:

$$M(OCOR)_2 + \overset{|}{\underset{|}{-C-}}_{Cl}^{Cl} \longrightarrow M\underset{OCOR}{\overset{Cl}{\diagup}} + \overset{|}{\underset{OCOR}{-C-}}$$

$$M(SR)_2 + \overset{|}{\underset{|}{-C-}}_{Cl}^{Cl} \longrightarrow M\underset{SR}{\overset{Cl}{\diagup}} + \overset{|}{\underset{SR}{-C-}}$$

It is understood that these stabilizer compounds can also react with any HCl formed before or subsequent to their incorporation into the polymer,

$$M(SR)_2 + HCl \longrightarrow M\underset{SR}{\overset{Cl}{\diagup}} + MSR$$

Mercaptides can also act as free radical quenchers.

It is further believed that the more modern stabilizers, organo-tin compounds (Table 2.3) and also the organophosphites, can act in this manner:

$$\sim\sim\!CH\!-\!CH_2\!\sim\sim + \underset{Bu}{\overset{Bu}{\diagdown}}Sn\underset{OCOR}{\overset{OCOR}{\diagup}} \longrightarrow \underset{Cl}{\overset{Bu}{-CH\!-\!CH_2-}}\!\sim\sim$$

$$\underset{Bu}{\overset{Cl}{\diagdown}}Sn\underset{OCOR}{\overset{OCOR}{\diagup}}$$

$$\sim\sim\!\!\underset{\cdot}{CH}\!-\!CH_2\!\sim\sim + P(OR)_3 \longrightarrow \sim\sim\!\!\underset{O=P(OR)_2 + RCl}{CH\!-\!CH_2}\!\sim\sim$$
$$+ \cdot Cl$$

Finally the disruption of conjugation in the polymer backbone chains can be achieved with mercaptides and maleate esters according to the following mechanism:

$$\sim\sim\!\!CH\!=\!CH\!\sim\sim + RSH \longrightarrow \sim\sim\!\!\underset{SR}{CH_2\!-\!CH}\!\sim\sim$$

$$\sim\sim\!\!CH\!=\!CH\!-\!CH\!=\!CH\!\sim\sim + \underset{CH}{\overset{CH}{\underset{\|}{}}}\!\!\underset{COOR\,(or\,M)}{\overset{COOR\,(or\,M)}{}}$$

COOR   COOR (or M)
(or M)

2.1.5 Synergistic stabilizer systems

The term synergism is used to describe the combined effect of two or more stabilizers, which is greater than the sum of the effects of the individual stabilizers used in isolation,

$$\text{Effect }(A + B) > \text{Effect } A + \text{Effect } B$$

With antioxidant systems, for instance, a combination of primary and secondary stabilizers are frequently used, especially in polyolefins, as they provide a very efficient stabilization (Fig 2.3).

Synergistic systems are even more widely used with polyvinyl chloride polymers. For instance, cadmium/barium and cadmium/barium/zinc stearates and laurates are notably good stabilizers and account for about 50% of all PVC stabilizers used commercially.

They are believed to function in the following manner[1]

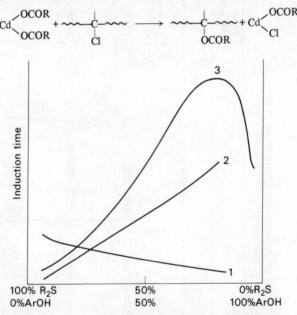

Fig. 2.3 Synergistic effects between primary and secondary stabilizers. Dependence on the induction period in the oxidation of polypropylene on the molecular fraction of di-decyl sulphide and 2,6-di-tert-octyl cresol. Total stabilizer concentration = 0·2 mol/kg at 200°C and oxygen pressure of 300 mm Hg. (After Neiman, M. B., *Ageing and Stabilization of Polymers*, Consultants Bureau (N.Y.), 1965, p. 30.)

Table 2.3 Typical stabilizers for vinylchloride polymers
(Total level of addition: approximately 0·5–2·0 p.h.p.)

Chemical name	Structure	Trade name and supplier	Comments
Lead stabilizers			
Basic lead carbonate	$PbO \cdot PbCO_3$	White lead (Associated Lead Manufacturers Ltd.)	Low cost and efficient. Toxic
Tribasic lead sulphate	$3PbO \cdot PbSO_4 \cdot H_2O$	TBLS (As above)	Low cost and efficient. Used in cable formulations.
Dibasic lead phosphite	$2PbO \cdot PbHPO_3 \cdot \frac{1}{2}H_2O$	DPLP (As above)	Good heat stabilizer, offers synergism in ageing stabilization and chelation towards metal impurities.

Mixed cadmium barium zinc stabilizers			
Coprecipitated cadmium/barium laurates	$Cd(OCOC_{11}H_{25})_2$ $Ba(OCOC_{11}H_{25})_2$	Mellite 101 (Allbright & Wilson (Mfg) Ltd.)	Susceptible to sulphide staining.
Barium/cadmium/zinc complexes	—	Nuostabes V1277 (Durham Raw Materials Ltd.)	
Organo-tin stabilizers			
Dibutyltin maleates	$\left[(C_4H_9)_2SnOCOCH=CH\ COO \right]_n$	Mellite 135 (Allbright & Wilson (Mfg) Ltd.)	Good high temperature stabilizer.
Dibutyltin, bis(lauryl) mercaptides	$(C_4H_9)_2Sn(SC_{12}H_{25})_2$	Mark A (Angus Chemicals)	Excellent for clear clear applications.

If used alone the Cd(OCOR)Cl would react with a further labile Cl atom in the polymer chain to form $CdCl_2$ which would, at this point, accelerate the HCl evolution process by a Friedel-Crafts alkylation mechanism.

In the presence of a barium carboxylate the formation of $CdCl_2$ is delayed considerably according to the equations below:

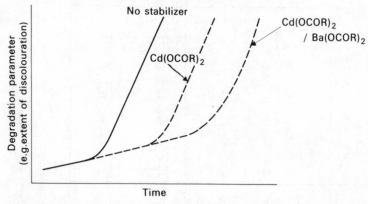

The function of the barium compound is, therefore, to regenerate cadmium carboxylate which has a strong stabilization power and to delay the formation of noxious $CdCl_2$. This would result in a large increase in induction time as indicated in Fig. 2.4[2].

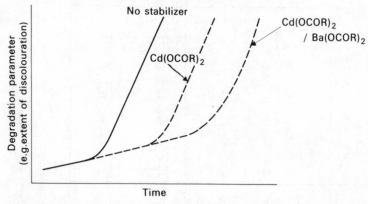

Fig. 2.4 Synergistic effects in stabilization of PVC by mixed metal carboxylates.

Even greater synergism is obtained by the addition of epoxy compounds, of which the more important are epoxidized soya bean oils, Fig. 2.5. It appears that the oxyrane rings aid the transfer of the chlorine to the barium carboxylate[1].

2.1.6 Evaluation of processing stabilizers

In the evaluation of additives in polymeric materials two types of objectives must be distinguished, one aimed at the elucidation of the

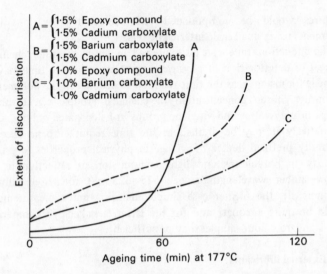

Fig. 2.5 Synergistic effects in PVC stabilization. (After Chevassus, F. and de Broutelles, R., *The Stabilization of Polyvinyl Chloride*, Edward Arnold, 1963.)

mechanism of the action of the additive and another concerned with a selection procedure and tailor making of the overall composition to achieve 'optimum' performance. Model systems would be used to achieve the first objective whereas for the latter, one would use formulations and procedures which are akin to those which would ultimately be used in production. Model studies on processing stabilizers involve the use of liquid (low molecular weight) compounds similar in structure to the polymer and/or the potential stabilizer under examination. The individual components of the reaction mixtures subjected to various heat and/or oxygen attack treatments would be identified, separated and estimated quantitatively by established chemical analysis techniques, such as infra-red spectroscopy, gas-liquid chromatography etc. There are difficulties, however, in interpreting these results in terms of 'true' polymer processing stabilization since the procedure does not allow for intervening factors, such as diffusibility, compatibility with the polymer and interactions with other additives.

Hence to obtain a more realistic assessment and to achieve the second objective it is more usual to carry out the evaluation using the actual polymer and stabilizer systems intended for ultimate use. The mixtures would be compounded and subjected to various heat treatments using small scale internal mixers.

The induction time can be expressed in terms of either chemical changes or deterioration of physical properties. Induction time may be defined, for instance, as the time required to produce in the polymer an arbitrarily chosen concentration of chemical groups, e.g. carbonyl groups in polyolefins and vinylene groups in halogenated polymers. Or alternatively it may be defined as the time required to produce an arbitrarily prefixed degree of change in physical properties, e.g. melt viscosity in polyolefins and light transmission or reflection in the yellow—amber wavelength range (575—625 nm) for vinyl chloride polymers. In the former case simple melt-flow-index measurements would be quite adequate and for the latter, a naked-eye comparison with standard colour samples may be sufficient.

2.2 External lubricants

General principles   The use of external lubricants, in order to decrease frictional forces and to reduce wear at the interface of members rubbing against one another, has been an established procedure for centuries. An active interest in the subject was in fact taken by Leonardo da Vinci who first laid down the foundations of the theories of friction and wear.

There are three methods of lubrications normally used.

(a) Fluid (or hydrodynamic) lubrication, which is effected by means of 'oils' or 'greases' (mainly hydrocarbons or silicones). A relatively thick layer of 'non-corrosive' fluid (in excess of 10^{-4} cm or 10 000 molecules)[3] is placed at the interface of the sliding or moving members, thereby preventing them from coming into direct contact. Frictional forces can be reduced in this way to the level of the internal forces (viscosity) of the fluid, and at the same time, by preventing the concentration of stresses at the asperities of irregularities at the surface of the sliding components, the formation of harmful debris is prevented.

(b) Boundary lubrication. This type of lubrication differs apparently from the above with respect to the thickness of lubricant film formed at the interface, and takes place under conditions which prevent sufficiently thick films of lubricants from being formed, i.e. when extraordinarily high loads are exerted at the interface. Stationary molecular layers of lubricants are present at the tip of the asperities of irregularities on the surface of the rubbing members and exert a lubrication action by virtue of their low cohesive energy (or strength).

It is, therefore, desirable that lubricants in boundary lubrication have a 'chemical' affinity for the rubbing components so that a permanent layer at the interface can be formed. Effective lubricants in the case of metal components are long-chain fatty acids, metal soaps and alcohols.

Since the effectiveness of alcohols, and acids in particular, depends on the nature of the metal surface and on their mutual reactivity, some form of chemical reaction between lubricant and the surface of the metal components is desirable. It is important to note, however, that the effectiveness of boundary lubricants (or that of the effective lubricant in the case where chemical reactions take place at the interface) falls off rapidly at temperatures around their melting point[4].

(c) Solid layer lubrication. This type of lubrication consists of coating the surfaces of the rubbing members with lamellar solids, e.g. graphite, molybdenum disulphide, etc., and are normally used under extremely high friction conditions, where very high temperatures are developed and which would render both fluid and boundary lubrication ineffective.

2.2.1 External lubrication of plastics to aid processing

There are three major friction problems encountered in plastics processing:

(a) Interparticle friction between polymer powders or granules which impairs their 'free-flowing' characteristics, hence creating difficulties in conveying operations, such as their feeding into processing equipment.

(b) Friction between the polymer melt and the metal surfaces of the processing equipment which can seriously impair the flow of the melt, and therefore produce undesirable effects, such as low output and poor surface quality of finished product.

(c) Friction in finishing operations, e.g. printing and packaging.

To alleviate the above difficulties two methods are available at present: solid layer lubrication for powder particles, and boundary lubrication for the melt/metal and solid polymer/metal interfaces.

External lubricants used to aid processing are similar to those used for conventional boundary lubrication of metal surfaces, i.e. high molecular weight fatty acids, alcohols, amines, amides and metal soaps containing 12—18 C atoms in the backbone chain. More recently waxes based on high molecular weight hydrocarbons (MW = 500—5000) have also been used.

The mechanism of melt lubrication is similar to that operating in conventional boundary lubrication, i.e. the polar groups of the lubricant molecules face the surface of the metal and form a stationary layer by virtue of strong absorptive forces.

The layer of lubricant film consists of oriented molecules which can be considered to be chemically associated[5] (Fig. 2.6), and therefore prevents their becoming solvated by mobile polymer molecules (see later).

This mechanism applies to all polymer systems, irrespective of whether they are highly polar or non-polar. More important than the final details of the structure of the lubricant, on the other hand, is the nature of the metal surface as it determines the intensity of the adsorptive forces which it exerts towards the polar groups of the lubricant molecules, and therefore its ability to form permanent boundary layers. As already explained the nature of the metal surface

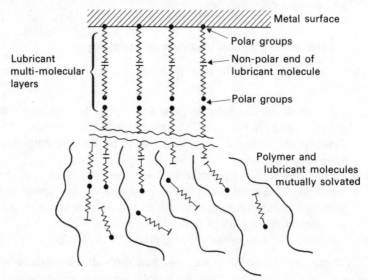

Fig. 2.6 Boundary layer of lubricant between metal surface and polymer melt.

can determine whether chemical reactions or physical adsorption can take place at the lubricant/metal interface. When the lubricant is an acid, lauric acid for instance[4], it will react with metals such as copper, cadmium and zinc but not with silver, aluminium, nickel, chromium, etc. Bowden and Tabor found that the coefficient of friction between metal and lubricant is considerably lower when chemical reactions take place. It is understood, however, that because the lubricant is imbedded in the melt during compounding or in the early stages of the melt mixing in processing, its effectiveness depends on:

(a) its lack of compatibility with the polymer melt so that the boundary layer at the interface is not re-dissolved by the polymer;
(b) its diffusibility in the melt so that it can easily migrate to the interface.

The compatibility of lubricants with polymer melts is generally low (see later) and consequently the formation of a boundary layer can be achieved with very small levels of incorporation. Migration rates of lubricants to the interface, on the other hand, are quite low owing to (a) their fairly high molecular weight, (b) low concentrations in the total system and (c) the relatively mild acceleration to which they are subjected in conventional processing. Thus there is a delay time for the lubricant to become effective in the processing of plastics, but in practice this may not present any real problem because of the inevitable delays for all the processing variables to adjust themselves to the 'steady state' conditions required for normal production. To facilitate handling operations of packaging films, 'slip' additives which have similar chemical constitution to external lubricants, are normally used.

Examples of external lubricants and slip additives are given in Table 2.4.

2.3 Internal lubricants and processing aids (high temperature plasticizers)

The function of internal lubricants and processing aids is to bring about plasticization at high temperature (see Fig. 3.9) so that the fusion rate of polymer particles when rubbing against one another, as in the first stage of processing, is increased and the deformability of the resulting melt is increased.

Ideally internal lubricants and processing aids should be compatible with the polymer only at high temperatures (i.e in the region of the rubber/melt transitions). If the compatibility is retained at low temperatures (i.e. in the region of the glass/rubber transitions) the lubricant or the processing aid should either be used in small concentrations or have a 'spectrum of relaxation times' in this temperature range (see Chapter 3) similar to that of the polymer in order to prevent plasticization taking place also at low temperatures.

In general internal lubricants are chemically similar to external lubricants with the exception that they exert greater compatibility and do not migrate readily to the surface. Their use is practically restricted to rigid PVC formulations and the normal level of addition is normally between 1 and 2 parts.

Processing aids, which in the case of thermosetting compositions are also called 'flow promoters', are low molecular weight polymers or

Table 2.4 Typical commercial lubricants and processing aids available commercially[6]

Lubricant or processing aid	Formula	Polymer for which the additive is recommended	Comments
1 Monoglyceryl esters	O, RO Chain C 14–18	PVC	Internal lubricants
2 Steraric acid	O, HO Chain C 14 18		
3 Waxes	ester, amide groups, Chain C 16. 18, Chain C 16 18	PVC, polyolefins	Internal/ external lubricants. Slip additives
4 Hydrocarbon waxes		PVC	External/ internal lubricants.
5 Soaps	Metal stearates	PVC, polyolefins	External lubricants.
6 Acrylates	Terpolymers based on:— methyl acrylate-butadiene-styrene, and acrylonitrile-butadiene styrene.	PVC	Processing aids.

waxes (Table 2.4). In the case of thermoplastics they act in the same manner as internal lubricants, i.e. they plasticize the outer surfaces of polymer particles and ease their fusion, but can be used in greater concentrations (about 5 parts per hundred). In the case of thermo-setting systems the processing aid or flow promoter is not reactive normally and therefore reduces the rate of interactions of reactive groupings by a dilution effect. Hence easier processing may be derived mainly from the reduction in the rate at which the melt viscosity increases. At the same time the overall cross-linking density of the resulting thermoset product is reduced and thereby producing simultan-eously some internal* and external plasticization effects (Chapter 3, p. 45). The use of processing aids or high temperature plasticizers in thermoplastics is generally limited to rigid PVC formulations although small amounts are sometimes also added to high viscosity polymers, e.g. polymethyl methacrylate, polystyrene, cellulose acetate, etc. to produce the so called 'easy flow' moulding grades (Plate 1).

2.4 General comments on external and internal lubrication of polymers

Although, as already mentioned in Chapter 1, thermodynamic consider-ations of polymer solutions may be used as a guide to predict the limits of compatibility of lubricants, the relevant data on lubricants in common use are still not available. Hence the lack of scientific information on lubricants, coupled with the very complex behaviour of polymer melts during processing, has left the whole subject of polymer lubrication in a state where selection and judgment is still made on a trial-and-error basis.

The normal rule used for the selection of lubricants is as follows:

(a) metal soaps, mainly stearates, possess low compatibility with all polymers and therefore act primarily as external lubricants,

(b) long chain fatty acids, alcohols and amides function as internal lubricants for polar polymers, e.g. PVC, polyamides, etc., but have relatively low compatibility with non-polar polymers, e.g. poly-olefins.

(c) long chain di-alkyl esters have medium compatibility with most polymers and can act both internally and externally, hence they are often used to obtain a balanced lubrication[6],

---

*Internal plasticization denotes a reduction in stiffness due to molecular structure effects.

A (a)

(b)

Plate 1 Effects of processing aids on high temperature fusion characteristics (A) and deformability (B) of rigid PVC.
(a) PVC without processing aids.
(b) PVC containing 10% Acryloid K – 120N.
(Reproduced from Rohm and Hass technical literature)

a)

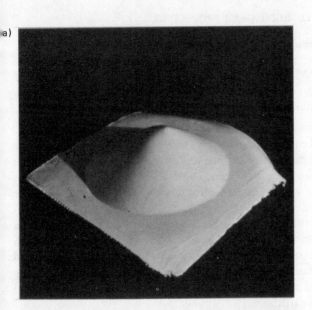

b)

(d) high molecular weight paraffin waxes have very low compatibility with polar polymers and act as external lubricants for polymers such as PVC. These are less widely used with polyolefins because of their high compatibility, which would therefore only provide internal lubrication. This is not normally required with polyolefins.

Special care must be taken in the selection of lubricants, especially when mixtures are considered, in order to balance the internal and external effects. Interactions with each other or with other additives present, e.g. stabilizers, could affect the compatibility limits and give rise to undesirable effects, such as 'plate-out' and over lubrications.

Plate-out is a term used to describe the deposition of residues from the melt on to the metal surfaces of the processing equipment, which may mar the surface finish of unsupported melts, such as extruded products or calendered sheets. These deposits may consist of accumulated lubricant at the interface, containing entrapped polymer and hard (e.g. pigment) particles.

Over lubrication, on the other hand, would result in lower outputs owing to excessive inter-particle and particle/metal slippage produced during the fusion stage. This will have the effect of lowering the melting rate of the polymer and reducing the thrust on the forward conveying action of continuous rotating Archimedian screws in extrusion and injection moulding.

Using an instrumented extruder Illmann[6] was able to obtain curves which represented over lubrication and under lubrication respectively (see Fig. 2.7).

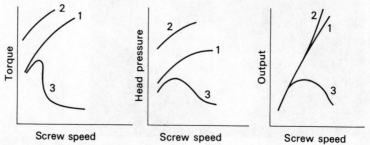

Fig. 2.7 1. Well balanced lubrication.
2. Lack of internal lubrication or overall lubrication insufficient.
3. Over lubrication.
(After Illman, G.[6])

The type of process and the range of temperatures and pressure to which the melt is subjected during processing, are other important factors that must be taken into account, since these may also affect the compatibility limits and the rate of lubricant migration.

2.5 Evaluation of processing aids and lubricants

Model studies on lubricants and processing aids for plastics have not been widely reported as yet.

Evaluations are normally carried out either with the aid of viscometers or on miniature compounding and processing equipment, such as the Brabender Plastograph or the RAPRA Variable Torque Rheometer.

Using capillary viscometers the effects of external lubricants can be measured by determining the value of the critical shear stress for the onset of 'wall slippage' or 'cold flow' and by measuring the slip velocity at higher shear stresses. According to Lupton and Regester[7] such information can be obtained from plots of $Q/\pi a^3$ v.s. $1/a$, where Q is the volumetric flow rate through the die and a is the capillary radius. In absence of slip $Q/\pi a^3$ is independent of a.

Using a cone-and-plate viscometer, internal lubrication and high temperature plasticization by means of processing aids can be assessed by monitoring the changes of normal forces and viscosity. Accurate estimates of external lubrication may be more difficult to obtain using the latter technique, since wall slippage is primarily a high shear rate phenomenon and presently available cone-and-plate equipment, such as the Weissenberg Rheogoniometer, may not be capable of operation at the required shear rate range. With the aid of miniature internal mixers, on the other hand, it is possible to measure the solvation rate of internal lubricants by recording the time for homogenization of the melt mixture, but it is less amenable to studies of external lubrication. Furthermore the latter may interfere with the solvation and fusion process, hence great care and experience is needed in the interpretation of the results.

2.6 Thixotropic agents or anti-sag additives

For the processing of very fluid polymeric systems, e.g. pastes, liquid resins, very low viscosity melts, etc. it is often desirable that the behaviour is either pseudoplastic or thixotropic (Fig 3.9).

The incorporation of insoluble additives with a very large surface area ($\cong$ 200 m^2/g), commonly known as thixotropic or anti-sag agents, form a continuous network through the fluid matrix and therefore hinder the Brownian movements of molecules or any other microscopic particles present. Such movements are hindered by adsorptive forces acting over the very large interfacial area of contact. Hence the most effective systems are those where there are hydrogen-bonding possibilities between the surface of the thixotropic agent and the fluid polymeric matrix.

With increasing duration of shear the network structure can be broken down by overcoming the adsorptive forces of the matrix molecules in the vicinity of the interface, hence the effective viscosity may be reduced (thixotropic effect). A similar effect can be obtained by increasing the shear rate (pseudo-plastic effect).

In both cases the ordered matrix structure at the interface can be re-established upon the termination of the shear owing to reduction in molecular motions. Chemical reactions between the additive and the polymer system must be prevented in order to facilitate the breakdown of the matrix structure at the interface. This is particularly important in the case of thermosetting resin systems. In these cases the thixotropic agent is often deliberately coated, or its surface chemically modified to reduce the possibility of interfacial chemical reactions. An example of such a modification is the neutralization of the basic hydroxyl groups on the surface of chrysotile asbestos fibres[8] when used as a thixotropic agent for unsaturated polyester resins. The thixotropic efficiency of chrysotile asbestos would, in this case, be affected by the reactions between the terminal acidic groups of the polyester resin and the outer hydroxyl groups on the surface of the asbestos. Sometimes the affinity between the thixotropic agent and the fluid polymer matrix is deliberately enhanced in order to achieve a more stable network structure. For instance, certain bentonite clays derived from a sodium aluminium silicate are rendered more effective in organic polymer systems by replacing part of the surface cations with alkyl amines or 'onium' compounds[9].

Typical thixotropic and anti-sag agents are shown in Table 2.5 below.

Table 2.5 Examples of thixotropic and anti-sag agents available commercially

Name	Comments	Recommended uses
Colloidal asbestos	Chrysotile variety fibre dia. $\cong 300$Å	Hydrophylic systems e.g. nylons, polyesters, etc.
Magnesium oxide	–	Polyester resins
Bentonite clays	Plate like structure $\cong 1\ \mu$m x 30Å	Polyester resins, PVC pastes
Silica flour	Particulate structure $\cong 0{\cdot}002 - 0{\cdot}07\ \mu$m surface area $\cong$ 200–400 m^2/g	Polyester resins, PVC pastes

References

1. *Encyclopaedia of Polymer Science and Technology*, Vol. 12, Interscience, 1970, pp. 744-7.
2. CHEVASSUS, F. and BRANTELLES DE, R., *The Stabilisation of Polyvinyl Chloride*, Edward Arnold, 1963, p. 170.
3. SCHNURMAN, R., Friction and Wear, *Wear*, 5, 1962, p. 1.
4. BOWDEN, F. P. and TABOR, D., *The Friction and Lubrication of Solids*, Oxford University Press, 1950.
5. AKHMATO, A. S., *Molecular Physics of Boundary Friction*, Israel Program for Scientific Translations, 1966, p. 285.
6. ILLMAN, G., *SPE Journal*, 23, No. 6, June, 1967, p. 72.
7. LUPTON, J. M. and REGESTER, J. W., *Poly. Sci. & Eng.*, October, 1965.
8. GUMP, J. A., *Modern Plastics Encyclopaedia*, McGraw-Hill, 1968, p. 427.
9. U.S. Patent, 2,531,427.

3 ADDITIVES WHICH MODIFY THE MECHANICAL PROPERTIES

Plastics materials derive their mechanical properties from the inter-actions of structural parameters such as molecular weight, branches in molecular chains, intrinsic flexibility of polymer chains, crystallinity and degree of order of molecular chains, crystal arrangements and intermolecular forces.

3.1 Deformation behaviour of plastics

The complex molecular structure of polymers does not allow plastics materials to behave as ideal materials in either the solid state or the melt state. The deformational behaviour of plastics (thermoplastics in particular) can be considered as a hybrid of the two ideal states, i.e. elastic solids and Newtonian liquids, and for this reason they are often referred to as elasto-viscous or visco-elastic materials.

It is possible to illustrate the visco-elastic behaviour of plastics by means of models, namely a spring to represent the elastic component and a dash-pot to represent the viscous or Newtonian component (Fig. 3.1). The deformation (strain) of such a model under the influence of a constant stress σ is given by:

$$\epsilon_{(t)} = \frac{\sigma}{E_0} + \sigma\psi_{(t)} + \frac{\sigma \cdot t}{\eta_0} \tag{3.1}$$

Hence the deformability or compliance ($J = \epsilon/\sigma$) becomes:

$$J_{(t)} = J_0 + \psi_{(t)} + t/\eta_0 \tag{3.2}$$

where

(i) E_0 represents the elastic modulus component whose magnitude is related to intermolecular forces between molecular chains.

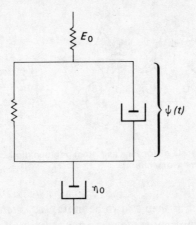

Fig. 3.1 Idealized behaviour of plastics.

(ii) η_0 represents the Newtonian viscosity component whose magnitude is related to molecular weight, chain branches and intrinsic chain flexibilities.

(iii) $\psi_{(t)}$ represents the 'creep function' and is associated with the strain of the middle spring and dash-pot in parallel (Voigt element). It is related to molecular parameters ($E_{(v)}$ and λ), according to the expression

$$\psi_{(t)} = \frac{1}{E_{(v)}} (1 - e^{-t/\lambda}) \tag{3.3}$$

In this case, $E_{(v)}$ is related to both bond rotational energy and polarity of groupings in the molecular chains and λ is a parameter related mainly to chain flexibility and free-spacings around segments of the molecular chains. λ is normally called the 'retardation time'*, and t is the time over which the deformation is considered.

The model in Fig. 3.1 can be used to describe the behaviour of thermoplastics over a wide range of temperatures. At very low temperatures it is the intermolecular forces which determine predominantly the behaviour of plastics. If the time of duration of the deformation is also very low, the last two terms of the equation can be neglected and the behaviour becomes equivalent to that of an elastic

---

*Normally λ is called the 'retardation spectrum' to indicate that it is a function capable of assuming a very large number of values distributed around a central value.

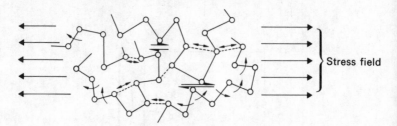

Fig. 3.2 Bond angle distortions and atom or group displacements when polymers are stressed at low temperatures.

material, which is then said to be in its *glassy state*. In other words, at very low temperatures the internal energy of the system is not sufficient to allow rotation of chains, hence the deformations are small and occur through small elastic distortions of bond angles in the molecular chains and through extremely small displacements of atoms or groups from different chains (Fig. 3.2).

At very high temperatures, assuming that the polymer does not undergo any thermal decomposition, the molecules possess a high level of internal energy and can, therefore, undergo Brownian movements. On application of stresses, molecular chains uncoil and recoil instantaneously, dissipating the applied energy into kinetic energy (i.e. viscous flow). The behaviour under these conditions approximates that of a Newtonian liquid, and is determined primarily by the value of η_0 (i.e. the contribution from chain length, chain branches and intrinsic chain flexibility, hence the first and second term of equation (3.2) can be neglected).

At some intermediate levels of temperature conditions can be met where there will be equal contributions from $E_{(v)}$ (bond rotational energy, polarity and λ (chain flexibility), when the deformational behaviour becomes similar to that of rubbery material, hence the polymer is said to be in its *rubbery state*. Deformations occur via uncoiling of chains and the applied energy is stored as a result of the increased entropy (i.e. $\Delta U_{stor.} \cong T\Delta S$). The internal energy is not sufficient to cause rapid recoiling of these chains, so if the time of duration of the stresses is relatively short, dissipation of the stored evergy by viscous flow does not take place.

Diagrammatically we can express the deformability of thermoplastics as a function of the ratios t/λ, where λ decreases with increasing temperature owing to increased chain mobility acquired by polymer molecules (Fig. 3.3).

In practice the Newtonian or viscous state (at least for thermoplastics) is rarely experienced owing to the onset of degradation reactions which prevents them from being processed at very high temperatures and, for economic reasons, they are rarely processed at sufficiently low shear rates. The transitional states, on the other hand, can cover a wide range of temperatures, making the behaviour of plastics in these states an important one in practice.

The glass/rubber and rubber/melt transitions may be considered as hybrids of the glassy and rubbery states and of the rubbery and melt states respectively. Hence the following models can be used to represent their behaviour (Figs. 3.4 and 3.5).

The standard linear solid model implies that the energy is stored as the result of externally applied stresses and the deformations recover on removal of these. The Jeffrey model, on the other hand, implies that part of the energy is dissipated by viscous flow and therefore the deformations are only partially recoverable.

Crystalline thermoplastics display a somewhat more complex behaviour than the one described above and the transitional states can span

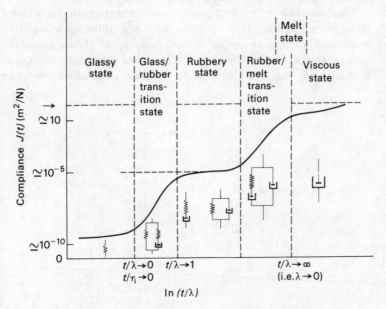

Fig. 3.3 Idealized deformational spectrum of thermoplastics.

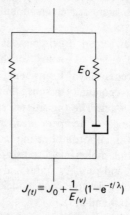

$$J_{(t)} \equiv J_0 + \frac{1}{E_{(v)}}(1-e^{-t/\lambda})$$

Fig. 3.4 'Standard linear solid' model representing the glass/rubber transitional state.

over an even wider range of temperatures. Other transitions can also occur within each transitional state as a result of the segmental motions, in the amorphous phase at lower temperatures, and in the crystalline phase at higher temperatures.

Thermosetting resins have considerably lower molecular weights than thermoplastics. This structure does not allow an orientation mechanism to operate owing to rapid recoiling of chains, hence the rubbery state is absent. Therefore thermosetting resins can only exist in a glassy state and in a viscous state. In those cases where molecular

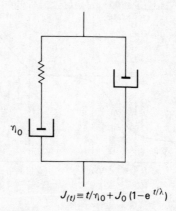

$$J_{(t)} \equiv t/\eta_0 + J_0 (1-e^{t/\lambda})$$

Fig. 3.5 Jeffrey model representing the rubber/melt transitional state.

weight is somewhat higher (e.g. certain epoxy resins) a transitional glass/melt state can exist and may occupy a moderate temperature range.

Thermoset plastics, on the other hand, being highly cross-linked can only exist in a glassy state and in a glass/rubber transitional state.

3.2 Fracture properties of plastics

The fracture behaviour of plastics is often described in terms of stress at break (strength) and energy to fracture (toughness).

Deformability parameters (e.g. modulus and compliance), strength and toughness can be obtained from load/deformation curves taken up to the point of fracture (Fig. 3.6).

According to Griffith, to produce fractured surfaces the externally applied energy must reach the value of the surface energy of the material:

$$U_F \geqslant 2\gamma \qquad (3.4)$$

where γ is the surface energy and the factor 2 indicates that two new surfaces are created during fracture.

If any amount of energy (U_D) is absorbed through molecular orientation prior to or dissipated during the fracture process, then we can re-write the fracture conditions as

$$U_F = 2\gamma + U_D \qquad (3.5)$$

Energy can also be dissipated however as heat, and sound. Energy absorbed through molecular orientation accounts for the high toughness of many plastics in the glass/rubber transitional state and rubbery state (Fig. 3.6).

The term 'elastic energy' used in Fig. 3.6 refers to the energy stored in the material as bond strain energy before fracture takes place. Possibly this energy is released as 'surface energy' during fracture propagation.

The term 'plastically stored energy' is used, on the other hand, to describe that amount of energy which remains stored in the material as internal- and free-energy (in this case 'orientation energy') after fracture.

The foregoing arguments imply that there are two basic principles underlying fracture processes:

(i) Energy is stored in the straining of chemical bonds and by molecular orientation. The fracture process begins by formation of

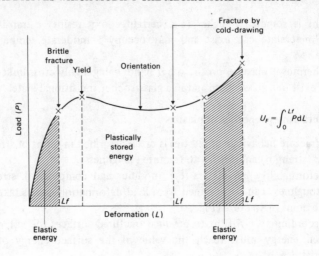

Fig. 3.6 Typical load/deformation curve for thermoplastics in their glass/rubber transitional state.

unstable cracks resulting when the intermolecular forces are overcome either in the weakest domains of the bulk of the material (e.g. regions of low molecular weight species) or in boundary regions where the highest level of bond-strain energy is stored (e.g. areas of highly cross-linked materials in the case of thermosets).

(ii) Once a crack has been formed, or if an 'effective' crack is already present (e.g. in presence of microscopic inclusions poorly bonded to the matrix), fracture occurs when the stored energy at the apex of the crack exceeds the surface energy of the material together with the energy absorbed by further molecular orientations around the crack (e.g. crazes, etc.) (Plate 3, p. 96).

Catastrophic fracture in a stressed material can, therefore, be prevented by allowing the crack to spread in many directions, or by promoting the formation of multitude of fine short cracks. In other words, by forming a large surface area around the propagating crack, the stored elastic energy may be released before the crack spreads across the whole width and thickness of the material. If, at the same time, molecular orientation also takes place, even if it occurs only in

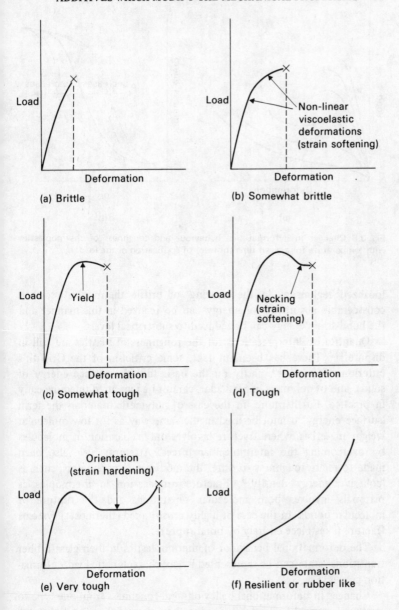

Fig. 3.7 Fracture behaviour of thermoplastics.
N.B. The fracture energy increases rapidly when molecular orientation begins to take place, i.e. from (c) to (f).

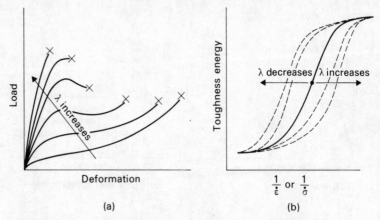

Fig. 3.8 Changes in deformational behaviour and toughness of thermoplastics with temperature for a given time (or rate) of application of the load.

localized regions as in the 'crazing' of brittle thermoplastics, then considerable absorption of energy can be realized in this manner and the bond strain energy can be reduced to sub-critical levels.

Quantitative interpretations for the toughness of plastics are still in dispute.[1,2] There has been, in fact, some criticism of the Griffith's criterion of fracture[4] partly on the basis that the surface energy of solids, and of polymers in particular, cannot be measured unequivocally in practice. Furthermore, in the case of polymeric materials the term 'surface energy' cannot be used in the same way as for low molecular weight materials, where fracture involves only separation of molecules by overcoming the intermolecular forces. Attempts have also been made to relate toughness to other thermodynamic parameters, such as 'cohesive energy density'[3]. Fracture propagation in thermoplastics materials involves both molecular separations and the rupture of molecular bonds. In the case of highly cross-linked (thermoset) systems fracture takes place entirely by bond rupture.

The deformational behaviour of thermoplastics in their glass/rubber transitional state can be represented by six characteristic load/deformation curves (Fig. 3.7).

Changes in deformational behaviour and toughness from one type to another is brought about by altering the value of the retardation (or relaxation) time function relative to the time (or rate) of application of the load. Decreasing the temperature has the effect of increasing λ and, therefore, promotes shifts towards glassy state behaviour. Vice versa, increasing the temperature promotes changes in behaviour towards that of the rubbery state (Fig. 3.8).

In the case of thermosets the deformational behaviour does not change very much with temperature and the behaviour is equivalent to that described by curves (a), (b) or (c) of Fig. 3.7. The large increase in toughness, achieveable with thermoplastics, e.g. types (e) and (f) of Fig. 3.7 and (b) of Fig. 3.8, can be attributed to the substantial storage of energy achieved through molecular orientation (at least for amorphous materials).

3.3 Principles for the modification of mechanical properties by additives

The main parameters of polymeric systems, $E_{(v)}$ and λ, previously described, can be altered by changing the environment of the polymer molecules i.e. by incorporation of soluble additives. As a result the position of the transitional states relative to the time (or rate) of the application of the load will change (Figs. 3.9 and 3.10).

As already mentioned in Chapter 1, there are three cases of additive polymer mixtures to be considered:

(1) In mixtures where polymer and additive are mutually compatible, the environment of chemical groupings in the polymer chains is altered by the surrounding additive molecule. In this case there will be a shift in the transitional states by an order of magnitude

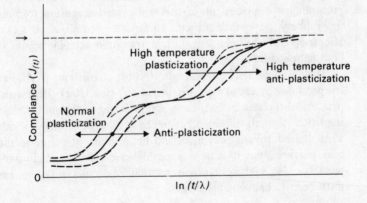

Fig. 3.9 Idealized changes in deformational behaviour by means of additives.

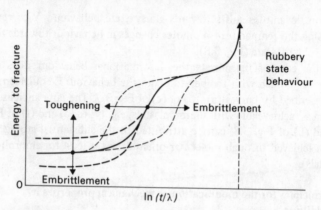

Fig. 3.10 Idealized changes in fracture behaviour of plastics in their glassy and glass/rubber transitional states by means of additives.

dependent on the relative effects which they have on each of the two parameters considered. That is to say that if the intermolecular forces are affected mostly then there will be a substantial shift in the glass/rubber transition. Whereas if the molecular chain rotations are affected mostly the shift will occur in the rubber/melt transition predominantly.

'Normal plasticization' is the term used to denote a shift in the glass/rubber transition towards the glassy state, i.e. the glass/rubber transition in presence of 'plasticizers' will occur at lower temperatures.

'High temperature plasticization' occurs, on the other hand, by promoting a displacement of the rubber/melt transition towards the rubbery state. It is difficult to separate completely the two effects and, therefore, some shift in transition actually occurs at both temperature ranges.

The displacement of the glass/rubber transition to higher temperatures is called 'anti-plasticization' (see later). High temperature anti-plasticization, i.e. a displacement in the rubber/melt transition towards higher temperatures, may be brought about when the additive is polymeric and has a similar structure to the base polymer, but it is of a much higher molecular weight and, possibly, also highly branched so that molecular rotations are restricted by chain entanglements.

(2) In mixtures where polymer and additive are incompatible two

phases are present. Such polymer mixtures are normally called 'composites'. The only important cases of 'composites' at present are those in which the additive is a mineral filler or any other solid material which does not undergo any transition over the ordinary temperature range for both application and processing of plastics materials.

Normally these inclusions are much more rigid than the plastics matrix and the deformational properties of composites are bound between the arithmetic mean and the reciprocal of the harmonic mean of the individual properties of the two phases (see later).

Such behaviour is better illustrated by referring to Fig. 3.11.

The glass/rubber and rubber melt transitions of composites correspond, therefore, to those of the polymer matrix, whereas the deformability function assumes lower values than the matrix over the whole spectrum of the deformational behaviour.

(3) In mixtures where the additive and polymer are partially compatible there will still be two phases present. In this case there will be some plasticization or anti-plasticization occurring in the interfacial regions. Such systems are normally called 'polymer blends' or 'allomers' and are becoming increasingly more important commercially owing to their high toughness. Normally both phases are polymeric and the blends can, therefore, exhibit each transition of the two individual phases (Fig. 3.12).

3.4 External plasticization and plasticizers

In previous discussions it was implied that the primary function of external plasticizers is to displace transitional regions to lower

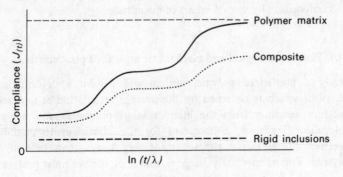

Fig. 3.11 Idealized behaviour of plastics composites.

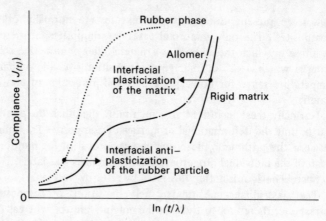

Fig. 3.12 Idealized behaviour of allomers.

temperatures. It was also said that it is possible to distinguish two ideal cases of plasticization:

(1) Plasticization which displaces to lower temperatures the rubber/melt transition, known as 'high temperature plasticization'. Plasticizers capable of producing this type of alteration in the deformational behaviour include processing aids and internal lubricants, whose sole function is, therefore, to assist the processing of the polymer without affecting its properties in the service temperature range.

(2) Plasticization which displaces the glass/rubber transitions into the range of temperatures below which the polymer is normally used in service in order to reduce the rigidity (modulus) of the polymeric composition by several orders of magnitude.

3.4.1 Phenomenological and mechanistic aspects of plasticization

Clearly a plasticized polymer system will exhibit a deformational behaviour which is governed by the average retardation or relaxation spectrum resulting from the interaction of plasticizer and polymer molecules. Hence, it is necessary for the plasticizer to exhibit a spectrum of relaxation times considerably lower than that of the polymer. This immediately suggests that a plasticizer must preferably be in the liquid state, or at least have a molecular weight considerably lower than that of the polymer.

In the case of crystalline polymers, plasticization is rather difficult unless the plasticizer has sufficient structural similarities with the polymer so that they can co-crystallize. In practice the majority of plasticizers have not such a structural similarity to commercial polymers hence they tend to be rejected, phase separation occurs and the plasticizer may exude out of the system.

If the crystalline polymer contains fairly large domains of amorphous material so that plasticization can effectively take place in these regions, a major shift in the glass/rubber transition can result. In practice, of all the commercially available crystalline plastics only polyamides, polyvinyl alcohol and polychlorotrifluoroethylene can be plasticized to an appreciable extent. The remainder tend to reject plasticizers at fairly low concentrations for the reasons set out above.

Because plasticizers act at molecular level the primary requirement is that polymer and plasticizer should be mutually soluble. This implies that solubility parameters for both polymer and plasticizer (δ_1 and δ_2) should be as close as possible, and, therefore, polarity and polarizability parameters should be well balanced. The fundamental concepts of solution thermodynamics, have been used, to some extent in polymer/plasticizer systems, but there is some evidence[1] that a 'true' solution is not always achieved with conventional plasticizers and that a dynamic equilibrium between plasticizer molecular aggregates and solvated polymer may be established instead. This hypothesis is also supported by certain anomalies which are observed with plasticized polyvinyl chloride polymers at both low and high levels of plasticizer. At low concentrations mechanical properties follow an inverse trend and at high concentrations two dielectric loss peaks[5] are experienced, which seems to suggest the existence of two phases. Consequently a scientific interpretation of plasticization phenomena, at present, can only be made in a qualitative manner. Usually the balance of polar polarizable non-polar components of the plasticizer molecule relative to the parent polymer determines their mutual solubility, whereas their relative relaxation characteristics are used to assess their ability to displace the glass/rubber transition temperature (plasticization efficiency). Note that polarity and polarizability can also affect the relative relaxation characteristics, hence compatibility and plasticization efficiency must be considered together.

3.4.2 Classification of plasticizers according to their compatibility and plasticization efficiency, and their evaluation

The three main types of intermolecular forces resulting from the non-polar, polar and polarizable constituents of polymer molecules can be subdivided into dispersive, dipolar, and inductive forces. These forces determine the solubility parameters of molecules and therefore compatibility will result if the balance of these forces in the monomeric units of the polymer and plasticizer molecules is similar. Hence most polymers will dissolve in their monomeric or telemeric species. In general, plasticizers which are totally compatible with the polymer are called 'primary plasticizers'. Since it is difficult, in practice, to obtain plasticizers which can meet all requirements (Section 3.4.3) and it is not usually necessary (with plastics) to produce excessively high plasticized polymer systems, since this would produce excessively soft, paste like materials, any plasticizer capable of exhibiting compatibility up to a ratio of 1:1 with the polymer can be included in this classification.

Any plasticizer with considerably lower compatibility, i.e. approximately up to 1:3 ratio with the polymer is classified as a 'secondary plasticizer'. When rather highly plasticized formulations are required, secondary plasticizers are normally used in mixtures with primary plasticizers.

Finally an additive which is not compatible with the polymer in ratios of at least 1:20, but exhibits a much higher compatibility with primary or secondary plasticizers so that it can be used in limited amounts for plasticization purposes, is known as a 'plasticizer extender'.

A plasticizer extender is often used in plasticized formulations for special reasons, e.g. to confer flame retardancy, anti-static properties, etc. It is understood that in the above classification an additive which may act as a primary plasticizer for one type of polymer may only be used as an extender in other polymers.

Examples of primary plasticizers for the following series:

> Polyamides, cellulose acetate, polyvinyl chloride, polystyrene and polyethylene

are respectively:

> N-ethyl o,p-toluene sulphonamide, tricresyl phosphate, di-octyl phthalate, hydrogenated terphenyls and hydrocarbon waxes.

If these plasticizers were listed in the reverse order, on the other hand, they would be incompatible with the polymers and could therefore only be used as plasticizer extenders.

The effect of the balance of polar, polarizable and non-polar constituents in plasticizer molecules is better illustrated with reference to polyvinyl chloride (Table 3.1). In this case, the balance in the plasticizer molecule is achieved by a suitable combination of ester (polar) groups, benzene (polarizable) groups and paraffinic (non-polar) groups[6].

Table 3.1. Balance of polar, polarizable and non-polar constituents in PVC plasticizers.

	Ester groups	Benzene groups	Methylene groups	Plasticizer type	Example
1	√	√		Primary	Tricresyl (TCP) phosphate
2	√		√	Secondary	Di-octyl (DOA) adipate
3	√	√	√	Primary	Di-octyl (DOP) phthalate
4		√		Extenders	Benzyl naphthalene
5		√	√	Extenders	Triphenyl ethane
6			√	Unsuitable as plasticizer	Octadecane

Finally, plasticizers can be classified on the basis of their plasticization efficiency. From the practical point of view plasticization efficiency is assessed by evaluating the low temperature flexibility of the polymer system in specially devised standard tests, e.g. bend brittleness test[7], cold flex test[8] and modulus measurements over a range of temperatures.

In Britain it is common to express the efficiency of a plasticizer for PVC in terms of a percentage plasticizer required to produce a plasticized composition which exhibits a certain pre-specified low

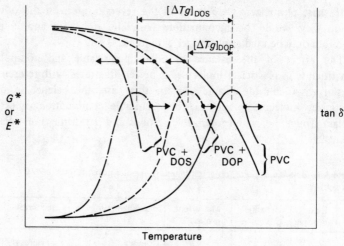

Fig. 3.13 Plasticization efficiency of DOS relative to that of DOP in plasticized PVC compounds. (Idealized curves)

$$\text{Plasticization efficiency} = \frac{[\Delta Tg/C] \ DOS}{[\Delta Tg/C] \ DOP}$$

where C = plasticizer concentration.

temperature modulus. In other parts of Europe, on the other hand, plasticizer efficiency is often expressed relative to that of a standard plasticizer (e.g. DOP for PVC formulations).

The temperature at which a torque of $5 \cdot 7 \times 10^{-2}$ Nm is required to produce a $90°$ rotation in a standard rectangular specimen is measured. This is known as the T_3 value. The efficiency of a plasticizer is expressed, therefore, as the ratio of its T_3 value to that of DOP when used at equal levels.

Plasticization efficiency can, however, be expressed more accurately in terms of lowering of glass transition temperatures as obtained from dynamic modulus measurements[9]. It is still possible to use a common plasticizer, e.g. DOP, as a standard and express the plasticization efficiency of plasticizers as the ratio of the respective decreases of glass transition temperatures (Fig. 3.13).

According to the extent to which the glass transition temperature of the base polymer is lowered, external plasticizers can be further divided into 'general purpose' and 'low temperature' plasticizers.

In the case of polyvinyl chloride, since the standard plasticizer is a 'general purpose' type, a 'low temperature' plasticizer will have a plasticization efficiency (as defined in Fig. 3.13) greater than 1.

Normally one single parameter, such as the lowering of glass/rubber transition, is not sufficient to characterize plasticization efficiency since it neglects the magnitude of the effects on modulus below this region. This is particularly important for the case of low temperature plasticizers (e.g. DOS in PVC) which tend to have a more pronounced effect on intermolecular forces, by a dipole screening mechanism, than on chain flexibility. Hence a modulus reduction ratio, G(polymer)/G(plasticized composition) determined over a wide range of temperatures may constitute a better alternative.

3.4.3 Classification of plasticizers according to their solvating power and migration characteristics

The previous classifications have been made on the basis of functional requirements, so that a particular type or combination of plasticizers could be selected for compatibility reasons and optimum level of plasticizer decided on plasticization efficiency grounds.

There are, however, two other incidental requirements that must be satisfied in selection of plasticizers, namely:

(a) they should 'solvate' the polymer at an economic rate at the compounding stage, and
(b) they should not exude out of the system through volatilization blooming or bleeding during the service life of the plasticized polymer.

Solvation and exudation rates are determined by the diffusibility of plasticizers through the polymer and are, therefore, dependent on both solubility parameters (intermolecular forces relationship) and molecular weight, but the latter may play an even more important role.

Another classification is therefore possible and plasticizers can be further divided into 'fast solvating' (or easy gellation) plasticizers and 'non-migratory' plasticizers.

There is obviously an inverse correlation between solvation rate and ease of migration of plasticizers since one process is the reverse of the other.

Frissel[10] has ranked plasticizers for PVC according to their minimum fluxing (or gellation) temperature in 'plastisol' formulations, and studied the effects of molecular weight and polarity by altering respectively the length of aliphatic chains and the nature of chemical groupings (e.g. ethers, amides, etc.) in the plasticizer molecule.

Solvation rates were found to increase with polarity and to decrease with molecular weight of the plasticizer.

Similar conclusions were also reached much earlier by Doolittle[11] for the solvation action of solvents on cellulose nitrate.

Regarding the migration rates of plasticizers, one must consider the effect of the environment. In the absence of a liquid environment, migration of the plasticizer can occur by volatilization (at high temperatures) and exudation or blooming (at lower temperatures). In the presence of a liquid or solid environment, on the other hand, the migration rate of the plasticizer depends to a large extent on the relative differences in solubility parameters between polymer, plasticizer and environment.

3.4.4 Commercial plasticizers

In Table 3.2 a few examples of plasticizers for common plastics materials are given and listed on the basis of the preceding classifications. For more detailed information the interested reader is referred to *Modern Plastics Encyclopaedia* (McGraw-Hill) and to *Encyclopaedia of Polymer Science and Technology* (Interscience).

3.5 Plasticizers anomalies and anti-plasticization

It has been recognized for a long time[12] that plasticizers for PVC show anomalies at low concentrations, and that the critical concentration (concentration above which the plasticizer exhibits normal behaviour) is inversely proportional to plasticization efficiency (Fig. 3.14).

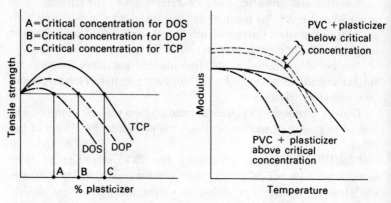

A = Critical concentration for DOS
B = Critical concentration for DOP
C = Critical concentration for TCP

PVC + plasticizer below critical concentration

PVC + plasticizer above critical concentration

Fig. 3.14 Mechanical properties of PVC as a function of plasticizer concentration.

More recent investigations[13] have shown however that this phenomenon is exhibited by most polymers possessing a rather polar structure and high chain stiffness (e.g. aromatic rings in backbone chains). The theories put forward are somewhat controversial insofar as most polymers that exhibit this behaviour are potentially crystallizable,

Table 3.2 Examples of commercial plasticizers for plastics materials

Plasticizer	Polymer	Plasticizer type
Di-octyl phthalate (DOP)	Polyvinyl chloride and copolymers	General purpose, primary plasticizer
Tricresyl phosphate (TCP)	Polyvinyl chloride and copolymers	Flame retardant, primary plasticizer
Tricresyl phosphate (TCP)	Cellulose nitrate	Flame retardant, primary plasticizer
Tricresyl phosphate (TCP)	Cellulose acetate	Flame retardant, primary plasticizer
Di-octyl adipate (DOA)	Polyvinyl chloride, cellulose acetate butyrate	Low temperature plasticizer
Di-octyl sebacate (DOS)	Polyvinyl chloride, cellulose acetate butyrate	Secondary plasticizer
Adipic acid polyesters (MW = 1500–3000)	Polyvinyl chloride	Non-migratory secondary plasticizer
Sebacic acid polyesters (MW = 1500–3000)	Polyvinyl chloride	Non-migratory secondary plasticizer
Chlorinated paraffins (%Cl = 40–70) (MW = 600–1000)	Most polymers	Flame retardant, plasticizer extenders
Bi- and terphenyls (also hydrogenated)	Aromatic polyesters	Various
N-ethyl o,p.toluene sulphonamide	Polyamides	General purpose primary plasticizer
Sulphonamide-formaldehyde resins	Polyamides	Non-migratory secondary plasticizers

and it is thought that the anti-plasticization effect could, in fact, be due to some crystallization of the polymer as the result of increased mobility of the polymer chains.

The fact that such materials retain their transparency and no real indication of crystalline domains have been detected by X-ray diffraction[14] except for PVC, it is believed that the anti-plasticization phenomenon is a real one and therefore the effect must be achieved as a result of increases in intermolecular forces and in reductions in the freedom of chains to undergo rotations and uncoiling.

The subject of anti-plasticization is still in its infancy and has not, therefore, received commercial exploitation, but the principle has been demonstrated and the relationship between polymer and anti-plasticizer molecular structure has been established.

It appears that in addition to polarity and chain stiffness in both polymer and anti-plasticizer molecules, it is essential that the additive be a molecule relatively free of steric hindrance so that efficient intermolecular forces transfer can take place between polymer chains.

Steric factors in bulky molecules would, in fact, create excessive internal voidage between polymer chains and reduce the overall intermolecular forces, hence producing a normal plasticization effect.

Examples of anti-plasticizers which have been investigated are given in Table 3.3.

Table 3.3

Anti-plasticizer	Polymer
Chlorinated terphenyls (%Cl = 40—60)	Bis-phenyl A — polysulphone ethers, polycarbonates
Polystyrene glycol (%MW = 500)	Bis-phenol A — polysulphone ethers, polycarbonates
Triethylene glycol ester of hydrogenated abietic acid	Bis-phenol A — polysulphone ethers, polycarbonates

3.6 Reinforcement and composites

The term reinforcement is normally used to denote the increase in modulus and in strength obtained when a second phase, intrinsically much stiffer and stronger than the base polymer, is dispersed in the polymer matrix. Hence the principles of reinforcement are based on the ancient concept of combining different systems to 'average out' the properties that each would exhibit in isolation.

This concept is particularly valuable in the context of composites since the properties of one of the elements (the reinforcing phase) cannot be exploited in isolation.

With plastics composites it is possible to distinguish two cases:

(1) The plastics material is used to encapsulate the reinforcing elements (long fibres) so that suitable shapes and structures can be fabricated.

The plastics material, in this case, is the minor constituent (20–50% v/v) and the composite is normally in the shape of laminates (*high performance reinforcement*).

(2) The properties of plastics are enhanced by incorporation of small amounts of short fibrous or particulate reinforcing fillers (5–25% v/v) so that ease of processing is retained. This type of reinforcement can be denoted as *low performance reinforcement* since the properties of the composite are nearer to those of the basic plastics material than to those of the reinforcing elements in isolation.

3.6.1 High performance composites (continuous fibres)

Unidirectional reinforcement

The case of longitudinal fibres has received greatest amount of attention by workers concerned with fundamental studies since it represents the simplest system and, therefore, it is more amenable to the development of the basic theories.

The emphasis, so far, has been placed on the prediction of properties of composites from the intrinsic properties of the matrix and the reinforcing elements, and the assumptions are normally made that each constituent of the composite behaves as it would in isolation, i.e. interactions are neglected. Various theories are available and the approach is often quite different, but the one which is mostly used is based on the 'law of mixtures'. The longitudinal and transverse response of a simple unidirectional composite (Figs. 3.15 and 3.16) are predicted by assuming respectively a parallel and a series reaction to the applied load[15].

(a) *Longitudinal response*

The response of unidirectional (continuous) fibres composites can be simulated by a model containing flexible and rigid springs to represent

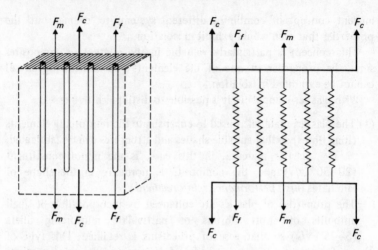

Fig. 3.15 Idealized behaviour of the longitudinal response of continuous unidirectional fibre composites (parallel reaction)[15].

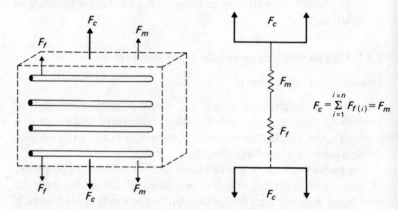

$$F_c = \sum_{i=1}^{i=n} F_{f(i)} = F_m$$

Fig. 3.16 Idealized behaviour of the transverse response of continuous unidirectional fibre composites (series reaction).

respectively the behaviour of the matrix and of reinforcing fibres:

$$F_c = \sum_{i=1}^{i=n} F_{f(i)} + F_m \qquad (3.11)$$

where: (a) the subscripts c, f and m refer to the composite, fibre and matrix respectively, (b) n = number of fibres, and (c) F = forces acting on each element.

This model assumes: (a) isostrain conditions, that is to say the strain across the composite is uniform, ($\epsilon_c = \epsilon_f = \epsilon_m$), (b) elastic behaviour of matrix and fibres, (c) the Poisson ratio of the matrix is equal to that of the fibres, so that debonding does not occur as a result of differential volume changes between the two phases.

Under these conditions the force components can be translated into stress terms by taking into account the relative cross-sectional areas of fibres and matrix respectively. Therefore, since $F = \sigma \cdot A$, where A is the cross sectional area, then

$$A_c \sigma_c = nA_f \sigma_f + A_m \sigma_m \qquad (3.6)$$

Also the volume of the composite (V_c) is the sum of the volume of the matrix (V_m) and that of the fibres (nV_f); where n = number of fibres V_f = volume of individual fibres

$$V_c = nV_f + V_m \qquad (3.7)$$

From this we can define 'volume fractions' for both fibres and matrix

$$\phi_m = \frac{V_m}{V_c}; \quad \phi_f = \frac{nV_f}{V_c}$$

from which we now obtain $\phi_m = 1 - \phi_f$.

Now, since $V = l \cdot A$, and the length of the fibres, matrix and composite is the same, we can substitute in equation (3.6) to obtain the relationship

$$\sigma_c = \phi_f \sigma_f + (1 - \phi_f) \sigma_m \qquad (3.9)$$

Dividing each term by the strain, $\epsilon_c = \epsilon_m = \epsilon_f$, we obtain an expression for the modulus of the composite in terms of volume fraction of the fibres, modulus of the matrix and modulus of the fibres·

$$E_c = \phi_f E_f + (1 - \phi_f) E_m \qquad (3.10)$$

(b) *Transverse response*

The properties of composites in the transverse direction to the fibres can be derived by means of a model consisting of springs in series. This model assumes:

(a) isostress conditions, i.e. $\sigma_c = \sigma_m = \sigma_f$
(b) elastic behaviour for the matrix and fibres
(c) the Poisson ratio of the matrix is equal to that of the fibres

For this model the strains terms are additive:

$$V_c \epsilon_c = n V_f \epsilon_f + V_m \epsilon_m \qquad (3.12)$$

hence

$$\epsilon_c = \phi_f \epsilon_f + (1 - \phi_f)\epsilon_m \qquad (3.13)$$

Dividing by the stress ($\sigma_c = \sigma_f = \sigma_m$) we can obtain an expression for the compliance and modulus of the composite:

$$J_{(c)} = \phi_f J_{(f)} + (1 - \phi_f)J_{(m)} \qquad (3.14)$$

and

$$\frac{1}{E_c} = \frac{\phi_f}{E_f} + \frac{(1 - \phi_f)}{E_m} \qquad (3.15)$$

or

$$E_c = \frac{E_f E_m}{\phi_f E_m + (1 - \phi_f)E_f} \qquad (3.16)$$

Similarly, assuming that no energy is dissipated from the system during the application of stresses (i.e. in absence of any dilation or interface slippage), we could have taken the stored potential energy/unit volume (strain energy) as being equal to the weighted algebraic sum of the energies stored by the two respective phases

$$U_c = \phi_f U_f + (1 - \phi_f)U_m \qquad (3.17)$$

and since $U = \tfrac{1}{2}\sigma \cdot \epsilon$, for the isostrain model we would obtain

$$\sigma_c = \phi_f \sigma_f + (1 - \phi_f)\sigma_m \qquad (3.18)$$

and for the isostress model

$$\epsilon_f = \phi_f \epsilon_f + (1 - \phi_f)\epsilon_m \qquad (3.19)$$

The longitudinal and transverse properties as a function of the volume of the fibres are represented in graphical form in Fig. 3.17.

Models could also be used to extend the treatment to cases where the polymer matrix behaves in linear-viscoelastic manner. The behaviour of such a composite would be similar to that of a standard linear solid (p. 48), in which the elastic component is modified by amounts proportional to modulus, volume fraction and orientation of the reinforcing fibres. This also implies that the time dependent

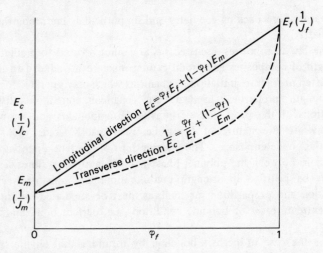

Fig. 3.17 Idealized behaviour of continuous unidirectional composites in the longitudinal and transverse direction.

behaviour of such composites is determined entirely by the matrix,

$$J_{m(t)}/J_{c(t)} = E_{c(t)}/E_{m(t)} = \text{constant} \qquad (3.20)$$

The ratio $E_{c(t)}/E_{m(t)}$ is known as the 'reinforcement factor' or 'modulus enhancement factor'.

In practical situations, where the matrix is normally a thermoset polymer (e.g. polyester or epoxide) the behaviour tends to be near-elastic up to the point where fibre debonding and matrix crazing takes place. After this point the behaviour becomes time dependent and is governed mainly by a matrix/fibre slippage mechanism.

With thermoplastics composites, the matrix is non-linear viscoelastic and therefore the modulus of the composite is also dependent on the strain level. If interfacial slippage takes place the reinforcement factor decreases with time and level of strain.

In practice, it has been found that the longitudinal thermoelastic properties (i.e. moduli, Poisson ratio, thermal conductivity, thermal expansion) for unidirectional reinforcement can also be represented by the simple law of mixtures. These simple models, however, lead to conservative estimates for the properties in the transverse direction. More comprehensive models show, in fact, that the transverse properties and shear moduli are sensitive to differences in Poisson ratios, fibre geometry, fibre packing geometry and, in particular, the properties of the matrix.

The law of mixtures discussed so far cannot be used to predict the strength of composites and modifications must be introduced to allow for different fracture initiation mechanisms which may operate.

Strength properties, in fact, are dependent on the statistical variations of the properties of individual components and are more sensitive to the nature of flaws, i.e. microcracks, voids, etc. than modulus or compliance. The assumption of perfect coupling (or adhesion) between the polymer phase and the reinforcing fibres can no longer be justified in strength considerations and possible fracture initiation and propagation mechanisms must be stipulated within the two extreme cases of bonding conditions, i.e. perfect bond and zero bond.

For the case of perfect bonding, by assuming that all the fibres fracture simultaneously, it is possible to stipulate a mechanism by which the sudden release of surface energy by the fibre fracture process causes rapid crack propagation through the matrix (Fig. 3.18), and the strength becomes:

$$\hat{\sigma}_c = \phi_f \hat{\sigma}_f + (1 - \phi_f)\hat{\sigma}'_m \qquad (3.21)$$

where $\hat{\sigma}'_m$ is the value of the stress for the matrix at strain equal to the fracture strain of the fibres.

Considering, now, the other extreme case where there is no adhesion between fibres and matrix, so that no load is carried by the fibres, when

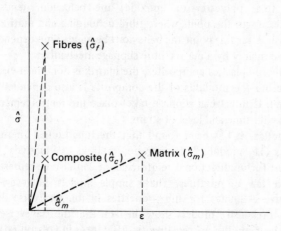

Fig. 3.18 Longitudinal strength of unidirectional composites for perfectly bonded fibres. 'Law of mixture' relationship.

failure of the matrix occurs there is a frictionless pull-out of the fibres and the strength of the composite becomes:

$$\hat{\sigma}_c = (1 - \phi_f)\hat{\sigma}_m \qquad (3.22)$$

In any intermediate situation, the strength of the composite lies between these two extremes and can be represented by the following equation:

$$\hat{\sigma}_c = k\phi_f\hat{\sigma}_f + (1 - \phi_f)\hat{\sigma}'_m \qquad (3.23)$$

where k = 'adhesion' factor or 'fibre utilisation efficiency' factor $(0 < k < 1)$.

This equation represents also the case where the fibres do not fail simultaneously, i.e. a proportion of these fibres fracture at stress levels below their average tensile strength. The presence of flaws on the surface of fibres can in fact cause statistical variations in their ultimate strength. The initial fracture of a proportion of the fibres does not result, however, in complete failure of the composite. As the two ends resulting from a fibre break attempt to pull away from each other the friction (shear stresses) at the interface (see later) can transfer stresses back into the fibre fragments (Fig. 3.19).

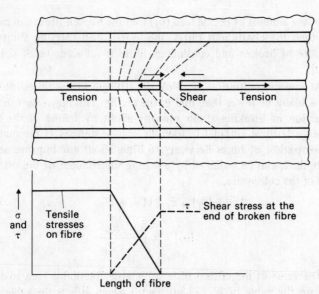

Fig. 3.19 Initial fibre fracture in unidirectional composites for three different lengths of fibre fragments. (After McCullough (Rev.) R. L., *Concepts of Fibre-Resin Composites*, Dekker, 1971, p. 49.)[15]

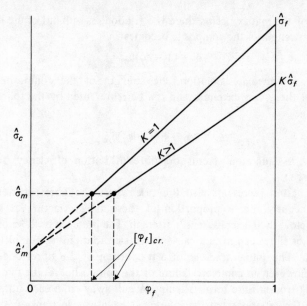

Fig. 3.20 Critical volumetric fractions in unidirectional composites.

Since the amount of tensile load (σ_f) that the broken fibres can carry is, less than the continuous fibres, the overall load carrying ability of a mixture of broken and continuous fibres is reduced, i.e. $K < 1$ (Fig. 3.20).

At low concentration of fibres the strength in the fibres' direction can actually be lower than that of the matrix. This may result from the reduction in breaking strain brought about by failure of the fibres. Hence a 'critical volumetric fraction' can be defined as the minimum concentration of fibres necessary to bring about any improvements in strength. It can be obtained by equating the strength of the matrix to that of the composite,

$$\hat{\sigma}_c = K [\phi_f]_{cr} \, \hat{\sigma}_f + (1 - \phi_f) \hat{\sigma}_m' \qquad (3.24)$$

$$\therefore \quad [\phi_f]_{cr} = \frac{\hat{\sigma}_c - \hat{\sigma}_m'}{K \sigma_f - \hat{\sigma}_m'} \qquad (3.25)$$

The value of the critical fibre volumetric fraction is seen to depend also on the value of K, i.e. any factor which affects the value of the 'effective' strength of the fibres in the composite.

In the transverse direction, once more, the law of mixture relationship can be used to describe the strength of composites with k values varying from 1, in the case of perfect adhesion, to 0 for no interfacial adhesion. This would be difficult to verify experimentally as there is no reliable means of measuring the transverse tensile strength of fibres. The enhancement of strength in the transverse direction, is normally very low and, in many cases is lower than the strength of the matrix ($\bar{\sigma}_m$). This is understandable since fracture can take place in the matrix plane where triaxial stresses are set up.

Angular properties

Simple models cannot be used any longer to predict the properties at any angle to the direction of the fibres because the isostress and isostrain conditions assumed in the models in Figs. 3.15 and 3.16 are too far from being realistic. The thermoelastic and viscoelastic properties in directions at an angle between zero and 90° to the fibres' direction will lie between the two bounds postulated in these two models.

It is possible to predict the properties of composites at any angle to the axial direction of the fibres by means of force vector diagrams (Fig 3.21).

For element 1, summing all the forces in direction T, one obtains

$$F_T + F_{12} \sin \theta + F_{21} \cos \theta - F_2 \cos \theta - F_1 \sin \theta = 0 \quad (3.26)$$

where

$$F_T = \sigma_T A; \quad F_2 = \sigma_2 A'; \quad F_1 = \sigma_1 A''; \quad F_{21} = \tau_{21} A'; \quad F_{12} = \tau_{12} A''$$

and

$$A' = A \cos \theta; \quad A'' = A \sin \theta$$

Hence

$$\sigma_T A + \tau_{12}(A \cos \theta)\sin \theta + \tau_{21}(A \sin \theta)\cos \theta - \sigma_2(A \cos \theta)\cos \theta$$

∴ since
$$- \sigma_1(A \sin \theta)\sin \theta = 0 \quad (3.27)$$

$$\tau_{12} = \tau_{21}, \quad \sigma_T = \sigma_2 \cos^2 \theta + \sigma_1 \sin^2 \theta - 2\tau_{12} \cos \theta \sin \theta \quad (3.28)$$

Summing all the forces in direction L using similar reasoning to above, one obtains

$$\tau_{LT} = \sigma_2 \sin \theta \cos \theta - \sigma_1 \sin \theta \cos \theta + \tau_{12} (\cos^2 \theta - \sin^2 \theta) \quad (3.29)$$

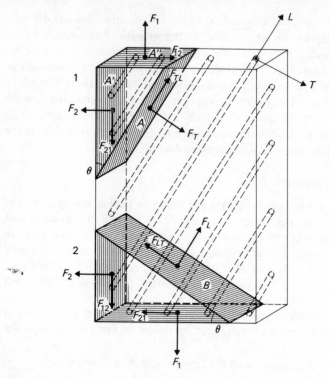

Fig. 3.21 Elements of a unidirectional composite, showing the resolution of forces.

For element 2, one obtains

$$\sigma_L = \sigma_2 \sin^2 \theta + \sigma_1 \cos^2 \theta + 2\tau_{12} \cos \theta \sin \theta \tag{3.30}$$

whereas σ_T and τ_{LT} will be the same as for element 1.

When θ is respectively zero and $90°$, the stresses acting on the two elements are σ_L, σ_T and τ_{LT}. We can derive similar expressions for the strain by considering the geometric relationships between the various displacement vectors in the two elements, and then define five elastic constants:

$$E_L = \frac{\sigma_L}{\epsilon_L}; \quad E_T = \frac{\sigma_T}{\epsilon_T}; \quad G_{LT} = \frac{\tau_{LT}}{\gamma_{LT}}; \quad \nu_{LT} = -\frac{\epsilon_T}{\epsilon_L} \quad \text{and} \quad \nu_{TL} = -\frac{\epsilon_L}{\epsilon_T}$$

which can be related to the fibres and matrix constants by the simple law of mixtures described in pp. 66–67.

To derive expressions for the case where θ is greater than zero and less than $90°$, one would have to consider the stress and strain tensor relationships. This treatment would be beyond the scope of this book; the interested reader will find, however, the necessary details in refs. 16, 17 and 18.

Assuming an elastic behaviour for both fibres and matrix, the Young's modulus for continuous unidirectional fibres at any angle from the direction of the fibres can be predicted by[17]:

$$\frac{E_L}{E_\theta} = \cos^4 \theta + \frac{E_L}{E_T} \sin^4 \theta + \left[\frac{E_L}{G_{LT}} - 2\nu_{LT} \right] \cos^2 \theta \sin^2 \theta \quad (3.31)$$

where E_θ = Young's modulus measured at angle θ to the direction of the fibres

E_L = Young's modulus in the direction of the fibres

E_T = Young's modulus measured in the perpendicular direction to the fibres

G_{LT} = Shear modulus in the direction parallel and perpendicular to the fibres

ν_{LT} = Major Poisson ratio

The strength properties at various angles from the fibre axis have been calculated by assuming that three failure modes can operate[19]:

Mode 1 The matrix deforms along the fibres until fracture takes place at right angles to the fibre axis.

Consider an element of the composite containing one single fibre at an angle to the direction of the applied force F_a (Fig. 3.22). In

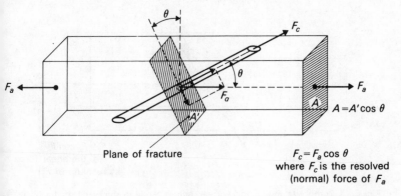

Plane of fracture

$F_c = F_a \cos \theta$
where F_c is the resolved (normal) force of F_a

Fig. 3.22 Composite model for a fracture mode at right angles to the fibres' axis.

such a model the force causing fracture is $\hat{F}_a$, which is greater than the fracture force $\hat{F}_c$ in the direction of the fibres.

$$\frac{\hat{F}_c}{A'} = [\hat{\sigma}_c]_{\,\prime\prime} = \hat{F}_a \cos\theta \Big/ \frac{A}{\cos\theta} = \frac{\hat{F}_a}{A} \cos^2\theta \qquad (3.32)$$

where

$$\frac{\hat{F}_a}{A} = [\hat{\sigma}_c]_{\theta}$$

Hence

$$[\hat{\sigma}_c]_{\theta} = \frac{[\hat{\sigma}_c]_{\,\prime\prime}}{\cos^2\theta} \qquad (3.33)$$

Mode 2 The composite can fail in the shear plane parallel to the fibre axis by a fibre/matrix debonding mechanism, and/or by shear failure of the matrix.

Consider a composite element containing one fibre at an angle θ to the axis of the applied stress (Fig. 3.23).

The force which causes fracture in this model is $\hat{F}_s$, hence

$$\hat{\tau}_m = \frac{F_s}{A''} = \hat{F}_a \cos\theta \Big/ \frac{A}{\sin\theta} = [\hat{\sigma}_c]_{\theta} \cdot \sin\theta \cos\theta \qquad (3.34)$$

that is

$$\hat{\tau}_m = [\hat{\sigma}_c]_{\theta} \sin\theta \cdot \cos\theta \qquad (3.35)$$

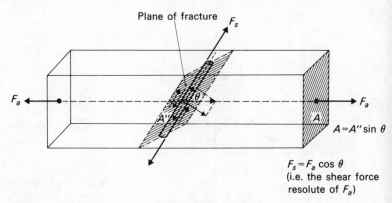

Plane of fracture F_s

F_a

F_a

$A = A'' \sin\theta$

$F_s = F_a \cos\theta$
(i.e. the shear force resolute of F_a)

Fig. 3.23 Composite model for a shear failure mode in the plane of the fibres' axis.

This type of failure occurs, therefore, under conditions where

$$\hat{\tau}_m / \sin \theta \, \cos \theta > \frac{[\hat{\sigma}_c]_{\prime\prime}}{\cos^2 \theta} \qquad (3.36)$$

or

$$\tan \theta > \frac{\hat{\tau}_m}{[\hat{\sigma}_c]_{\prime\prime}} \qquad (3.37)$$

The interfacial fibre/matrix strength and the shear strength of the matrix is much less than the strength of the composite in the direction of the fibres, hence this failure mechanism will operate when the angle between the applied stresses and the direction of the fibres is quite small, and the strength of the composite becomes

$$[\hat{\sigma}_c]_\theta = \frac{\hat{\tau}_m}{\sin \theta \, \cos \theta} \qquad (3.38)$$

Mode 3 Failure of the composite occurs by virtue of the matrix failing in tension in the plane parallel to the fibre direction.

Taking a similar composite element (Fig. 3.24) the strength of the composite can be predicted by similar arguments to those made for failure modes 1 and 2.

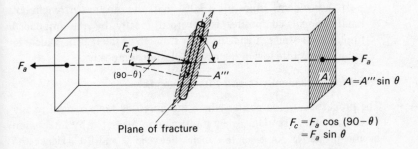

Fig. 3.24 Composite model for a tensile fracture mode in the plane parallel to the fibre.

$$\frac{\hat{F}_c}{A'''} = [\hat{\sigma}_c]_{90°} = \hat{F}_a \sin \theta \Big/ A/\sin \theta = \frac{\hat{F}_a}{A} \sin^2 \theta \qquad (3.39)$$

where

$$\frac{\hat{F}_a}{A} = [\hat{\sigma}_c]_\theta$$

and therefore

$$[\hat{\sigma}_c]_\theta = \frac{[\hat{\sigma}_c]_{90°}}{\sin^2 \theta} \qquad (3.40)$$

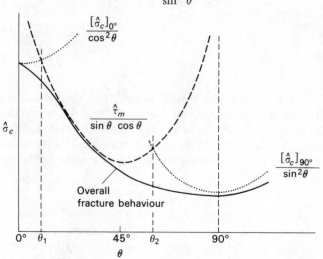

Fig. 3.25 Strength variations of unidirectional composites as a function of the fibre orientation angle.

It is evident from Fig. 3.25 that the angular strength of unidirectional composites is governed mostly by the interfacial fibre/matrix strength and the shear strength of the matrix, which, in general, are quite low in the case of plastics composites.

Properties of randomly-oriented-fibre composites

The properties of composites where the fibres are not arranged in any particular order (random orientation) can be predicted by integration of the properties over all the angles between 0 and 90°. Hence the modulus of a random fibre composite will be given by[20]

$$\langle E_\theta \rangle_c = \frac{\int_0^{\pi/2} [E_\theta]_c \, d\theta}{\int_0^{\pi/2} d\theta} \qquad (3.41)$$

The modulus $\langle E_\theta \rangle$ can be expressed by the simple law of mixture equation

$$\langle E_\theta \rangle = FE_f\phi_f + E_m(1 - \phi_f) \qquad (3.42)$$

where F represents a 'fibre efficiency' factor relative to that of the longitudinal direction of unidirectional fibre composites.

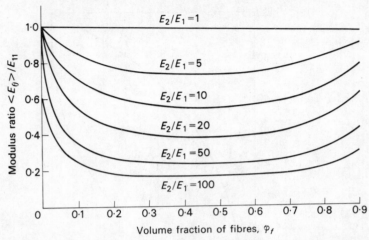

Fig. 3.26 Fibre efficiency factor F as a function of the volume fraction of fibres for different values of E_f/E_m. (After Nielson and Chen[20])

The fibre efficiency factor F is a function of the fibre volumetric fraction and of the ratio of the modulus of the fibres to that of the matrix[20], as shown in Fig. 3.26.

This analysis shows that for most plastics composites the fibre efficiency factor F varies from about $0 \cdot 15$ to $0 \cdot 60$, and that for random reinforcement a much lower reinforcing efficiency is to be expected than for unidirectional reinforcement. This is also shown in Fig. 3.27, where the ratio of the modulus of the composite to that of the matrix, (reinforcement factor) is plotted against volumetric fraction.

A similar approach could be used to predict strength properties:

$$\langle \hat{\sigma}_c \rangle_\theta = \frac{\int_0^{\theta_1} \frac{[\hat{\sigma}_c]_{0^\circ}}{\cos^2 \theta} \, d\theta}{\int_0^{\theta_1} d\theta} + \frac{\int_{\theta_1}^{\theta_2} \frac{\hat{\tau}_m}{\sin \theta \cos \theta} \, d\theta}{\int_{\theta_1}^{\theta_2} d\theta} + \frac{\int_{\theta_2}^{\pi/2} \frac{[\hat{\sigma}_c]_{90^\circ}}{\sin^2 \theta} \, d\theta}{\int_{\theta_2}^{\pi/2} d\theta}$$

$$(3.43)$$

The true fracture envelope can be represented by an equation of the type

$$\frac{1}{[\hat{\sigma}_c]_\theta^2} = \left(\frac{\cos^4 \theta}{[\hat{\sigma}_c]_{0^\circ}^2} + \frac{\sin^2 \theta \cos^2 \theta}{\hat{\tau}_m^2} + \frac{\sin^4 \theta}{[\hat{\sigma}_c]_{90^\circ}^2} \right) \qquad (3.44)$$

which can be averaged out to give an estimate of the strength of randomly oriented fibre composites.

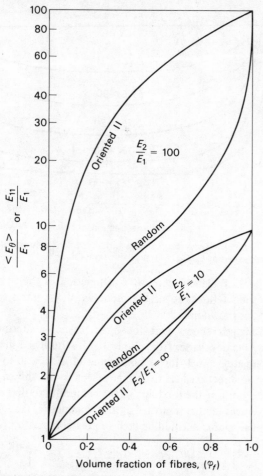

Fig. 3.27 Reinforcement factor of composites as a function of volumetric fraction and E_f/E_m ratio for unidirectional and random reinforcement composites. (After Nielson and Chen[20]).

3.6.2 Low performance reinforcement (short fibres composites)

When short fibres are used to reinforce plastice matrices neither of the conditions stipulated for continuous fibres reinforcement apply owing to fibre ends effects. Only a proportion of the fibre length may be carrying the maximum stress as shown in Fig. 3.28.

The two ends of each fibre are subjected to shear stresses, owing to the differential strain set up between the matrix separating the fibre

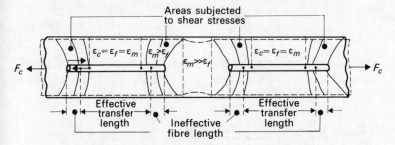

Fig. 3.28 Strain distribution in unidirectional short fibre composites.

ends and the matrix surrounding the fibres along their length. The fibre can be considered to have an 'ineffective fibre length'. The stress transfer at the fibre end may be related to the interfacial shear stresses in the manner shown in Fig. 3.29.

When the forces on the fibre ends balance exactly the reaction forces produced at the interface, we have

$$F_f = \overline{F}_{m/f}$$

hence

$$[\sigma_f]_e \frac{\pi d^2}{4} = \pi d \frac{a}{2} \bar{\tau}_{m/f} \tag{3.45}$$

and

$$a = \frac{[\sigma_f]_e d}{2\bar{\tau}_{m/f}} \tag{3.46}$$

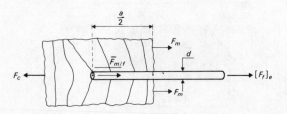

Fig. 3.29 Relationship between ineffective fibre length and interfacial shear stress.

or

$$[\sigma_f]_e = \frac{2\bar{\tau}_{m/f}\,a}{d} \qquad (3.47)$$

where a is the length of the fibres subjected to shear stresses and $[\sigma_f]_e$ is the value of the normal stress along the fibre ends. From this we can

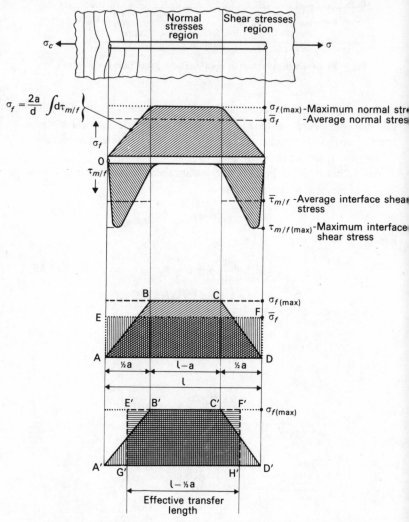

Fig. 3.30 Stress distribution along discontinuous fibres and significance of 'effective transfer length'.

calculate the average value of the normal stress σ_f acting over the entire fibre length and the correspondent value of the length of the fibres which carries the maximum normal stress.

The latter is known as the 'effective transfer length' and its significance is shown in Fig. 3.30.

$$\text{Area}_{(ABCD)} = \text{Area}_{(AEFD)}$$

∴

$$\therefore \quad \sigma_{f(\max)} \left[\frac{(l-a)+l}{2} \right] = \bar{\sigma}_f \cdot l \tag{3.48}$$

thus

$$\bar{\sigma}_f = \sigma_{\max} \left(1 - \frac{a}{2l} \right), \tag{3.49}$$

since

$$\bar{\sigma}_c = \phi_f \bar{\sigma}_f + (1 - \phi_f)\sigma_m \tag{3.50}$$

then

$$\sigma_c = \phi_f \sigma_{f(\max)} \left(1 - \frac{a}{2l} \right) + (1 - \phi_f)\sigma_m \tag{3.51}$$

Also

$$\text{Area}_{(A'B'C'D')} = \text{Area}_{(G'E'F'H')}$$

∴

$$\sigma_{f(\max)} \left(l - \tfrac{1}{2}a \right) = \bar{\sigma}_f \cdot l \tag{3.52}$$

and

$$\bar{\sigma}_f = \sigma_{f(\max)} \left(1 - \frac{a}{2l} \right) \tag{3.53}$$

hence the equivalence of the two approaches.

Equation (3.51) cannot be used directly to predict the modulus because neither the strain in the matrix nor the strain along the fibre axis is constant. Halpin and Tsai[21] have, however, obtained the following relationship:

$$\frac{E_c}{E_m} = \frac{(1 + \xi\eta \, \phi_f)}{(1 - \eta \, \phi f)} \tag{3.54}$$

where

$$\eta = \frac{\left(\dfrac{E_f}{E_m} - 1\right)}{\left(\dfrac{E_f}{E_m} + \xi\right)} \quad \text{and} \quad \xi = \frac{2l}{d}$$

In the transverse direction the modulus of the composite is not very sensitive to fibre length, hence it can be calculated from the compliance equation on p. 68.

We can use however equation (3.51) to obtain directly an estimate of the strength. In fact if we replace a by its equivalent value l_c for the case where the interfacial shear stress becomes equal to the ultimate fibre/matrix interfacial shear strength, then we can write:

$$\hat{\sigma}_c = \phi_f \hat{\sigma}_f \left(1 - \frac{l_c}{2l}\right) + (1 - \phi_f)\hat{\sigma}'_m \tag{3.55}$$

where l_c is known as the 'critical fibre length'. If the fibre length is at least 20 times greater than l_c then the strength of discontinuous fibres composites approaches closely that of continuous fibres laminates. If, on the other hand, the length of the fibre is smaller than l_c the only stress transfer mechanism is by interfacial shear and the strength in the longitudinal direction becomes

$$\hat{\sigma}_c = 2\phi_f \frac{l}{d} \hat{\tau}_{m/f} + (1 - \phi_f)\hat{\sigma}'_m \tag{3.56}$$

By using similar arguments to those used for the case of continuous fibres composites one could attempt to predict also the strength at various angles from the fibre directions, and possibly also for randomly oriented fibres. For the latter the analysis, if at all possible, would be very involved and it is doubtful whether it would have any practical value. The additional parameters of random distribution of fibre length, fibre misalignments, irregularity of fibre diameter and degree of fiberization (as in the case of asbestos or cellulose composites), voidage etc., would create so many uncertainties which may invalidate any theoretical predictions[22].

What really emerges out of the above treatment is that for short fibre composites normally used for moulding applications it is the interfacial shear strength which plays the most prominent role in determining the enhancement of stiffness and strength, (compare predictions from equations (3.47) and (3.56) with practical results in Plate 2).

Plate 2 Effects of interfacial adhesion on fracture properties of plastics composites. (Author's unpublished work)

(a) Fractured surface of glass fibre filled Nylon 66. (Strong interfacial bonding promotes fibre breaking mode of failure.)

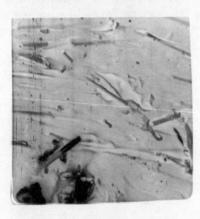

(b) Yielding of glass fibre filled polypropylene.

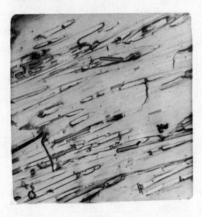

(c) Cold drawing of glass fibre filled polypropylene. (Poor interfacial bonding promotes fibre pull-out)

3.6.3 Particulate fillers composites

It is clear from the previous considerations that the reinforcement efficiency, especially with respect to strength properties, deteriorates considerably when the length of the reinforcing fibres fall below the critical fibre length. Hence a minimum enhancement of properties is to be expected when the aspect ratio of the fibres approaches unity and the filler action becomes primarily hydrodynamic[23]. Owing to the complex geometry of particulate fillers most of the treatments to date are largely empirical and attention has been focused more towards methods of improving interfacial bonding and to the mechanisms of fracture rather than to the intrinsic relationship between reinforcement efficiency and basic properties of the matrix and fillers.

The most widely used method for the prediction of thermoplastic properties is still the basic law of mixtures, where the composite properties are considered to be bound between the responses of the two basic models (i.e. in parallel and in series).

The bulk modulus and shear modulus properties are taken as the basic parameters for deriving stiffness relationships[24]:

$$\frac{1}{\dfrac{\phi_f}{K_f} + \dfrac{(1 - \phi_f)}{K_m}} \leqslant K_c \leqslant \phi_f K_f + (1 - \phi_f) K_m \tag{3.57}$$

and

$$\frac{1}{\dfrac{\phi_f}{G_f} + \dfrac{(1 - \phi_f)}{G_m}} \leqslant G_c \leqslant \phi_f G_f + (1 - \phi_f) G_m \tag{3.58}$$

The Young modulus will be derived by using the basic elasticity relationship $E = 2G(1 + \nu)$; $E = 3K(1 - 2\nu)$. These bounds are reasonably close when the difference between the moduli of the two phases are not too different (e.g. the case of polymer blends).

More realistic bounds for composites, where the difference in modulus between the two phases is substantial, have been obtained by Hashin and Strikman[24] and are expressed by:

$$K_{c(\text{lower bound})} = K_f + \frac{(1 - \phi_f)}{\dfrac{1}{K_m - K_f} + \dfrac{3\phi_f}{3K_f + G_f}} \tag{3.59}$$

$$K_{c(\text{upper bound})} = K_m + \frac{\phi_f}{\dfrac{1}{K_f - K_m} + \dfrac{3(1 - \phi_f)}{3K_m + 4G_m}} \tag{3.60}$$

In the case of rubbers, Van der Poel has solved the equations for the special case of a filler with a Poisson ratio of $0 \cdot 25$ assuming a value of $0 \cdot 5$ for the rubber. The variation of shear modulus with filler concentration was found to lie between the two bounds and could be expressed by the empirical equation:

$$\frac{G_c}{G_m} = 1 + \left[\frac{1 \cdot 25 \, \phi_f}{1 - 1 \cdot 28 \, \phi_f} \right]^2 \tag{3.61}$$

hence making the shear modulus of the composite completely independent of the properties of the filler. This confirms the earlier suggestion that provided the modulus of the filler is substantially greater than that of the matrix (e.g. $E_f/E_m > 20$) it is the interfacial bonding which plays the most important role.

The strength properties of particulate-filler composites are even more dependent on the geometry, size of the filler particles and to adhesion and interfacial contact of the two phases, than deformational properties.

Owing to stress concentration set up at the interface, cracks can be formed in the matrix at considerably lower stress levels than they would occur in absence of the filler inclusions. The severity of stress concentrations is greater with irregularly shaped particles and, since microcracks are also likely to be present at the interface due to poor wetting of the filler by the matrix, these are expected to give inferior strength reinforcement than regular (spherical) shaped fillers. It has been shown that stress concentrations increase at some distance from the filler surface, and if the concentration of rigid inclusion is high the matrix can actually be subjected to higher stress levels than the interface. A crack can, therefore, form in the matrix at modest stress levels and propagate towards the fillers. If the interfacial adhesion is poor the crack will spread around the surface of the filler (Plate 2), and consequently this will result in a decrease in strength (i.e. $\hat{\sigma}_c = (1 - \phi_f)\hat{\sigma}_m$). If the matrix is, on the other hand, firmly bound to the filler the cracks may be arrested and the stress will increase further until either the interfacial bond is broken or the filler particle itself fractures. Consequently the process of crack propagation will occur in sequences as shown below. (Fig. 3.31).

A more efficient reinforcement can also be achieved if a modulus gradient at the interface is established by the use of a cross-linked polymer. In this way stress concentrations are reduced and crack formation in the matrix will occur at much higher stress levels than the case where the modulus at the interface changes abruptly. No

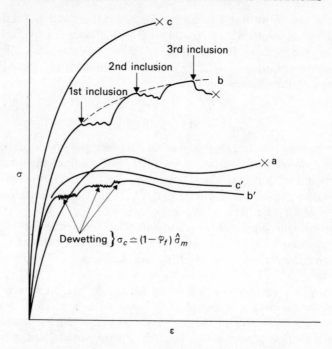

Fig. 3.31 Effects of inclusions in thermoplastics matrices.
a = matrix without filler; b = matrix containing 3 spherical inclusions (large size and strongly bound to matrix); c = matrix containing large quantities of spherical inclusions (strongly bound to matrix). Due to the very large number of inclusions the curve has become much smoother and b' and c' are same as b and c respectively but inclusions are poorly bonded to the matrix[26].

quantitative prediction, however, of the strength of particulate filler composites can yet be made. Particulate fillers are normally used in plastics compositions for the purpose of reducing formulation costs and to reduce thermal expansions or contractions. Approximate estimates for the reductions in thermal expansions may be obtained from equations (3.57), (3.59) and (3.60). Since

$$\alpha = \frac{1}{V} \cdot \frac{\Delta V}{\Delta T} \quad \text{and} \quad \frac{\Delta V}{V} \cong \epsilon_v, \quad \text{then} \quad K = \frac{P}{\epsilon_v} = \frac{P}{\alpha \cdot \Delta T}$$

(where α = volumetric thermal expansion coefficient, P = hydrostatic pressure). At constant pressure, therefore, the term K may be substituted for $1/\alpha\Delta T$.

3.6.4 Interphase regions of composites

It was demonstrated in the earlier sections that the properties of unidirectional composites at relatively large angles from the direction of fibres, and of randomly oriented fibre composites depend, to a very large extent, on the degree of interfacial bonding between fibres and matrix.

In practice a strong bond between fibres and polymer matrix is difficult to achieve owing to:

(a) Poor wettability of the polymer matrix over the very large surface area of the fibres, especially with the non-polar, high melt viscosity thermoplastics systems.

(b) The surface of most reinforcing fibres is hydrophylic and may be coated with multimolecular layers of water which may prevent physical or chemi-adsorption of the polymer matrix molecules. These water layers may be formed on subsequent ageing by diffusion of atmosphere moisture through the matrix

(c) A weak boundary layer at the interface may be formed by contaminants present on the surface of the fibres (e.g. lubricants, sizes, antistatic agents in the case of glass fibres) or by an entropic separation of low molecular weight additives from the polymer matrix.

The interfacial bond may be enhanced and the mechanical performance of composites improved by using suitable 'coupling agents' or adhesion promoters'.

Historically, glass fibres have been invariably used as reinforcing fillers and unsaturated polyesters and epoxide resins have been used almost exclusively for the matrix phase in laminated composite structures. Consequently most coupling agents have been developed specifically for these systems.

Five basic types of coupling agents are normally used, which consist of organic functional groupings capable of co-reacting with the polymeric matrix. These invariably contain vinyl and allylic groups for polyesters and amino groups for epoxides, together with inorganic or organo-inorganic groupings, which can exert a strong physical–chemical interaction with the hydroxyl groups of the glass surface. Examples of coupling agents are shown in Table 3.4.

In addition to chemically bridging the matrix to the glass surface, the function of the coupling agent is to reduce the rate at which water can accumulate at the interface and destroy the bond. To do so, in fact,

Table 3.4 Common coupling agents for glass fibres/thermoset resins laminates

Coupling agents	Recommended resin	Reaction with the glass surface
1 Vinyltrichloro silane $(CH_2=CH-SiCl_3)$	Polyesters	$CH=CH_2$ │ Si / │ \\ O O O ~~~~~~~ + 3 HCl Si Si glass O O O surface
2 Vinyltriethyoxy silane $(CH_2=CH-Si(OEt)_3)$	Polyesters	$CH=CH_2$ │ Si / │ \\ O O O ~~~~~~ + 3 EtOH Si Si Si glass /\\ /\\ /\\ surface
3 γ aminopropyl triethoxysilane $(NH_2-(CH_2)_3-Si(OEt)_3)$	Expoides	$(CH_2)_3NH_2$ │ Si / │ \\ O O O ~~~~~~ + 3 EtOH Si Si Si glass /\\ /\\ /\\ surface
4 Methacrylo-chromium complex $CH_2=C$⟨CH_3⟩ C⟨O⋯CrCl₂ / O–H / O–CrCl₂⟩	Polyesters	$CH_2=C$–CH_3 / C / O O Cr Cr O O O H O O ~~~~~ + 4 HCl Si Si Si Si glass /\\ /\\ /\\ /\\ surface
5 Allyltrichloro silane resorcinol $\left(CH_2=CH-CH_2-SiCl_2O \bigcirc OH \right)$	Universal	$CH_2=CH-CH_2$ │ Si / │ \\ O O O–⟨⟩ OH ~~~~ + 2 HCl Si Si Si glass /\\ /\\ /\\ surface

the water must first hydrolyse the −O−Si−O− bonds at the interface and displace both resin and coupling agent species.

With short fibres thermoplastics composites (e.g. moulded products), however, the use of conventional coupling agents does not seem to offer great advantages and the degree of properties enhancement depends mainly on the relative natural affinity of the matrix towards the fibres. Thus hydrophylic polymers (e.g. nylons, polysulphones, phenoxies) give better reinforcement than hydrophobic polymers (e.g. polyolefins, polystyrenes, etc.).

More successful results have recently been obtained by exploiting[26,27] the idea of producing a modulus gradient between the fibres and the matrix by means of cross-linked coatings whose network density gradually decreases towards the bulk of the matrix. Various techniques have been used to obtain a cross-linked coating strongly bonded to both fibres and the bulk of the matrix and, owing to their relatively low sensitivity to temperature changes and time factors, composites with superior properties at high temperatures and with better creep performance have been obtained.

Systems which can cross-link the matrix at the interface only have been patented and include the use of peroxides or chlorinated alicyclic compounds for polyolefin composites[26,27].

3.6.5 Evaluation of reinforcing fillers

The principles of deformation and strength of plastics and their modification by means of fillers discussed in preceeding sections have been illustrated mainly by reference to simple static (tensile) stresses.

A more comprehensive approach would have been necessary if these properties were to be dealt with in a design engineering context and the response to complex stress histories would have to be taken into account.

In the case of composites, in addition to the effects of fibre, matrix and fibre/matrix bonding properties already discussed, one would have to consider other influential factors such as voidage, fibre dispersion, etc. The lack of established theoretical procedures for designing with plastics and the complex interactions of the various components of composite materials make it impossible to predict with any degree of accuracy their performance without resorting to extensive experimental evaluations.

In practical evaluations of the reinforcing efficiency of fillers one is normally concerned with determining the enhancement in modulus and stress levels which cause failure. Generally, if the material undergoes fracture at low strain levels it is the ultimate stress which determines the failure conditions, whereas if the breaking strain is large (as in the case where fracture is preceded by yielding) the failure criterion is determined on a limiting strain basis, i.e. the stress which causes a pre-determined maximum deformation.

The high performance composites, i.e. those containing large amounts of a continuous reinforcing phase, are normally treated as elastic materials and the three elastic constants (or five constants if the material is anisotropic in one plane, p. 75) are normally measured on conventional, constant cross-head speed testing equipment using three point bending and torsion tests. Since failure normally takes place by shear in the interlayer regions, it is common to assess the relative enhancements in strength by measuring the interlaminar shear strength. This test is carried out by means of a three-point loading fixture using a span/specimen thickness ratio of $\simeq 4$ and recording the stress to failure.

If the value of the stress which causes a tensile failure through the composite cross-section is required, specially designed specimens will have to be used to enable measurements to be made on conventional testing equipment. The specimen which is most widely used for this purpose is the NOL-ring, made by filament winding and clamped by special jaws which support the specimen along the inner sides of the upper and lower sections of the ring, and transmits tensile stresses at the two edges, where failure is induced.

Short-fibre composites, especially those based on thermoplastics, have to be treated as viscoelastic materials and it is therefore essential to evaluate their time dependant characteristics.

Creep tests are now well established for such studies and can be carried out using tensile, bending and torsion deformational modes. The reinforcing efficiency is expressed in terms of 'fibre utilization efficiency factor' (p. 79) or 'modulus enhancement factor' (p. 69) at various time intervals and strain levels. Poor interfacial bonding normally gives rise to strain softening phenomena by virtue of secondary creep mechanisms caused by fibre/matrix dewetting and slippage.

To asess the enhancement in strength, the time dependancy is measured by normal tensile tests carried out at various straining rates and at constant loads, for long term evaluations.

It would also be necessary to take into account the effects of temperature (especially with thermoplastics) and the relative deterioration of interfacial bonding caused by water diffusion and capillary effects at the exposed fibre ends (at least for thermoset laminates).

In the author's laboratory and elsewhere a correlation has been obtained between dynamic losses and interfacial bonding. In the glass/rubber transition and rubbery states a 'poor' interfacial bonding tends to increase the damping coefficient as a result of fibre/matrix friction, whereas a 'good' bond may decrease appreciably the dynamic losses owing to a reduction in chain flexibility of adherent polymers layers. In this way one can obtain rather quickly qualitative predictions of the reinforcing efficiency of short-fibre composites at large strain and high temperatures.

3.7 Toughening by means of rigid and rubbery inclusions

From considerations in Section 3.4, it is evident that an appreciable increase in toughness can be achieved through plasticization owing to the greater freedom acquired by the molecular structure to undergo rotations and orientation. Plasticization, however, brings about a considerable reduction in the rigidity of the material, which may be undesirable. If the toughness of relatively brittle plastics is to be increased without sacrificing rigidity to any significant extent, the only suitable means of achieving this, is to produce two-phase (or multiple-phase) systems. The dispersed phase can provide suitable mechanisms for the effective release of the strain energy according to any of the principles illustrated in Section 3.2.

3.7.1 Toughening by means of fibrous fillers

As already explained, the introduction of high modulus fibrous fillers into plastics matrices will produce stress concentrations at the interface. These may result purely from the vast differences in modulus between the matrix and the filler or may occur at the tip of micro-cracks present at the interface resulting from poor wetting. Consequently fibrous composites invariably produce situations which are more conducive to crack propagation without the energy dissipating effects of yield-zones ahead of the crack. Nevertheless these systems can be tailor made in such a manner as to increase the overall fracture energy. The following three examples illustrate how this can be achieved:

(1) If failure begins in the matrix phase and the interfacial bond is strong, cracks can propagate through the fibres, hence toughening can only be achieved if intrinsically tough fibres, e.g. organic fibres are used.

   The fibre bundled structure of yarns or natural cellulosic fibrils, which would normally be used, would also cause further dissipation of energy through interfibrillar friction and, therefore, provide an efficient crack stopping effect.

For this reason and for their reactive sites on the surface they are mainly used with brittle thermoset polymers, such as phenolics, urea formaldehyde and melamine formaldehyde.

(2) The most likely failure mechanism in composites based on high modulus, high strength fibres, especially those containing discontinuous random fibres, is by a fibre/matrix interfacial shear failure. Although this mechanism prevents the composite from reaching its maximum potential strength, considerable expenditure of energy is realized by the pull-out action of the fibres giving rise to appreciable improvements in toughness. From Fig. 3.29.

$$U_{(\text{pull-out})} = \bar{F}_{m/f} \cdot l_c = \bar{\tau}_{m/f}\pi \, \text{d} \, l_c^2 \qquad (3.62)$$

Since the interfacial shear work is a function of fibre length the overall mechanical properties, e.g. stiffness, strength and toughness, are therefore optimum when long fibres are used and poor interfacial adhesion exists between fibres and matrix. A weak bond can also increase the toughness of composites in the cases where fracture begins in the matrix phase. Cook and Gordon[29] suggest, in fact, that when the crack approaches the fibre/matrix interface a crack stopping effect is established by virtue of a tensile failure at the interface. This failure spreads along the fibre's surface and reduces the intensity of the stresses at the crack tip. This type of toughening mechanism is likely to operate with glass fibre/polyester laminates.

(3) The most efficient way of toughening fibre reinforced composites is to establish a good interfacial bond, so that maximum strength is achieved, and to allow the fracture energy released by the fibres to be consumed through molecular motions (e.g. by using thermoplastics coatings in their rubbery or glass/rubber transition state). A good interfacial bond will ensure that the broken ends of the fibres retract, to some extent, after their fracture by virtue of the constraint imposed by the rigid matrix surrounding the flexible coatings on the fibres.

This mechanism has been suggested to operate in the case of polyesters and epoxide glass fibre composites, where the toughness has, to some extent, been attributed to the interfacial plasticization effects of the coupling agent or to entropic separation of low molecular weight species from the matrix[15].

3.7.2 Toughening by means of rubber particle inclusions

The most effective method of toughening brittle polymers is by means of rubber particles inclusions. This technique is widely used commercially to enhance the toughness of brittle thermoplastics (e.g. polystyrene, SAN, PVC). Various mechanisms have been put forward to explain the toughening action of rubber particles, but because of the complexity of the particles' geometry (Plate 3) and of the interfacial structure, quantitative predictions cannot be made with any degree of accuracy.

The following interpretations may, however, apply:

(1) The Poisson ratio of rubber approaches very closely the 0.5 value for undilatable and uncompressible materials, which is considerably higher than that of brittle thermoplastics ($0.35-0.40$). It is feasible, therefore, that if there is good adhesion between the rubber particles and the surrounding matrix, there will be considerable absorption of energy by the rubber particles owing to their undilatability (i.e. there will be an enormous increase in the bond strain energy of the rubber particles).

Consequently toughening is achieved by transferring large amounts of the externally applied energy from the matrix on to the rubber particles.

Since a polymer blend is, in effect, equivalent to a particulate composite material, a law of mixture relationship for the distribution of the stored energy between the matrix and the rubber particles may be used, (c.f. p. 68). This can be expressed in terms of energy per unit volumetric strain or energy per unit isotropic stresses.

$$\frac{U_c}{[\epsilon_v]_c} \equiv \frac{U_R \phi_R}{[\epsilon_v]_R} + \frac{U_m(1 - \phi_R)}{[\epsilon_v]_m} \left.\right\} \begin{array}{l}\text{For a parallel}\\ \text{response model}\end{array} \quad (3.63)$$

and

$$\frac{U_c}{(\sigma_c)} \equiv \frac{U_R \phi_R}{(\sigma_R)} + \frac{U_m(1 - \phi_R)}{(\sigma_m)} \left.\right\} \begin{array}{l}\text{For a series}\\ \text{response model}\end{array} \quad (3.64)$$

where U = stored energy

ϵ_v = volumetric strain

σ = isotropic stress

ϕ = volumetric fraction

subscripts c, R, m refer respectively to composite, rubber and matrix.

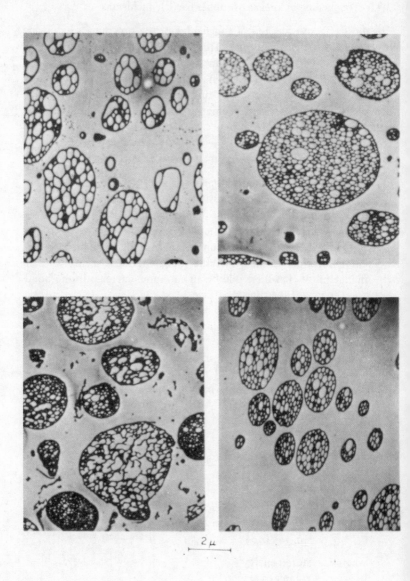

Plate 3. Electron photomicrographs of various rubber/polystyrene blends (After Keskkula, H., Polyblends and Composites, *J. App. Poly. Symposia*, **15**, p. 58).

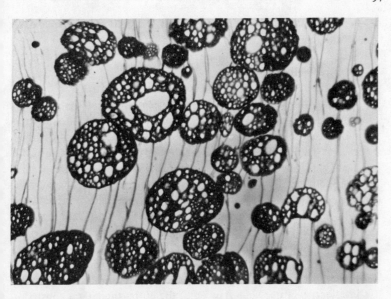

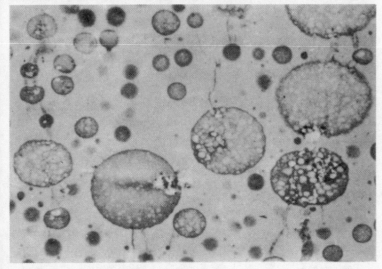

Plate 4 Crazing of thermoplastics/rubber blends.
(a) Crazing of toughened polystyrene. (After Bucknall C. B., Gotham K. V. and Vincent P. I.) [Reproduced from 'Polymer Science', Jenkins A. D. (ed.), North-Holland Publishing Co., Amsterdam (1972).]
(b) Crazing of ABS. (After Koichi Kato)
[Reproduced from *British Plastics*, **40**, November, 1967, pp. 119—20]

Hence the energy storage capacity of plastics/rubber blends is considerably increased, by the large values of the stored energy terms of the rubber phase, irrespective of the model used to describe their behaviour.

(2) The difference in moduli between rubber and matrix would cause, however, the setting up of stress concentrations at the interface, which would encourage the formation of many cracks around each rubber particle and release a considerable amount of the stored energy. Hence the larger the number of particles dispersed in the plastics matrix the greater the number of small cracks are formed and the larger the amount of energy absorbed by crack formation.

(3) With commercial polymer blends in the interfacial regions between rubber and matrix there is normally a layer of plasticized material. Tough failure by yielding and cold-drawing rather than brittle failure is, therefore, likely to occur in these interphase regions and propagate through the rigid matrix (Fig. 3.32 and Plate 4). The rubber particles and the matrix, at some distance away from the interface, will not cold-draw, and, therefore, the lateral constraint imposed by these promotes fibrillation (craze structure) in the interfacial region and dissipates large quantities of energy (Fig. 3.32). The whitening effect which is observed when fracturing toughened thermoplastics materials has, in fact, been attributed to the formation of interparticle fibrillation and crazing of the matrix, and has been confirmed by electron microscopy studies[31].

The modest increase in toughness obtained in the case of some commercial polymer blends (e.g. high impact polystyrene) may attributed to any of the following:

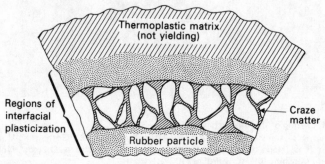

Fig. 3.32 Fibrillar structure formed in the interface regions between rubber particles and plastics matrix[30].

(a) The stresses acting on the matrix and on the rubber particles are not isotropic, and shear deformations may take place instead in the bulk of the rubber particles and in the interfacial regions. This is particularly so in the case where fracture occurs by slow crack propagation. If the interfacial bond is not particularly strong, these shear deformations would promote absorption of energy by the rubber particles by an extensional process rather than by dilation. The energy change associated with unconstrained elongation of rubber particles is normally very small. For an ideal rubber in fact $[\partial E/\partial L]_{T,V} \to 0$ and the potential energy barrier which must be overcome to obtain molecular orientation is also very small owing to very low intermolecular forces and bond rotational energy.

(b) The rubber particle inclusions do not exhibit an ideal behaviour, that is, the Poisson rate is actually less than $0\cdot5$. Hence $[\partial V/\partial L]_{T,P} > 0$ and, therefore, $[\partial E/\partial V]_{T,L}$ is not as large as it would have been if the ideal conditions of undilatability were achieved.

(c) Monomer residues or other impurities may act either as stress cracking agents, (i.e. they may supply large amounts of energy at the tip of a spreading crack through absorption and adsorption) or as nucleating centres for the formation of cracks.

(d) At low temperatures the rubber particles are brought into their glass/rubber transitional state. This transformation is accompanied by differential contractions between the matrix and the dispersed phase which may set up internal stresses in the interfacial regions (Fig. 3.33).

(e) The agglomeration of rubber particles may not provide an efficient crack stopping mechanism because of the intensive triaxial stresses field set up in the interstices of brittle matrix (Plate 4) by a mechanism similar to that discussed on p. 87.

3.7.3 Evaluation of the toughening efficiency of fibrous fillers and rubbery modifiers

The nature and magnitude of the stresses to which a component may be subjected in service is determined by its design. Engineering design criteria concerning fracture behaviour with plastics materials are still at infancy stage, thus it is often difficult to decide on the type of test

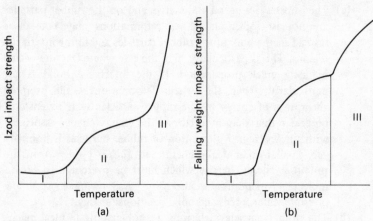

Fig. 3.33 Impact strength behaviour of high impact polystyrene sheets. (After Bucknall and Smith[32])

method that should be used to obtain appropriate data. Evaluations of the toughness of materials, however, inevitably involve studies of impact strength and fatigue resistance, using both notched and un-notched specimens.

It is customary to carry out impact strength measurements on pendulum instruments or by falling weight methods. The former technique normally uses notched specimens and provides an indication of the fracture resistance of materials in terms of energy per unit area of crack extension. It is widely used in so far as it gives a quick estimate of the brittle/tough transition temperature. Falling weight measurements, on the other hand, are much more time consuming since they require a statistical estimate of the likelihood of fracture at each energy input level. They are normally performed on sheets or pipes and will provide results which are more easily related to general service conditions for components without stress concentration features (e.g. sharp corners, holes, etc.). Thickness effects cannot be allowed for owing to the complexity of the propagation of stress waves within the specimen. Consequently this test method is not normally recommended for general evaluation purposes.

In recent years there have been numerous attempts to utilize fracture mechanics concepts for the evaluation of the toughness of plastics. Various procedures are available for this purpose but a full assessment of the relative merits of each has not yet been carried out.

Single-edge-notch specimens appear to be the most widely used and crack extension is brought about in tension or in bending. For a thorough evaluation the material would have to be tested over a wide range of temperatures under static (constant stress) and dynamic (cyclic stress) fatigue, and at various constant straining rates. Complications normally arise however, with respect to crack propagation owing to the spread of crazes, yield zones and fibre matrix debonding areas (for fibre-filled materials) around the crack tip. This suggests, therefore, that these tests are useful at least to determine the 'critical' conditions for the onset of crack propagation. These critical conditions may well constitute the safe design limits for short fibre reinforced thermosets.

In such test methods a measure of the toughness is given by the critical stress intensity factor, K_{IC}, where the subscript I denotes a crack opening mode normal to the crack plane. Measurements are normally made under the most severe conditions (i.e. plane strain).

Table 3.5 Fibres and fillers for plastics (characteristics and usage in commercial composites)

Fibres for high modulus and strength performance.

Chemical nature	Forms available	Density (d) (kg/dm³)	Specific modulus (E/d) (Nm/kg x 10^{-6})	Specific strength ($\hat{\sigma}_f/d$) (Nm/kg x 10^{-6})
(a) *Continuous fibres*				
Glass E type	Rovings, cloths	2·55	~30	~1·4
Glass S type	Non-woven unidirectional fabrics	2·49	~2·0	
Carbons	Filaments	2·00	200–300	0·7–1·0
Berilium	,,	1·84	~150–170	1·0–1·2
Boron	,,	2·59	~170	≅1·4
(b) *Discontinuous fibres (whiskers)*				
Silicon carbide	Fibres, wools (dia. ≅1–10 μm; l ≅ 20–1000 μm)	3·15	200–300	2–6
Silicon nitride	—	3·2	≅100	1–3
Aluminium oxide	Fibres, mats, papers	3·9	150–600	3–8
Berilium oxide	—	1·8	≅300	4–8
Asbestos	Fibres, mats, felts	2·5–3:0	≅40–60	0·5–1·0

The onset of crack growth under dynamic conditions is more difficult to assess and consequently it is more usual to follow the rate of crack propagation at various stress (or stress intensity factors K_I) levels. High toughness would be denoted by a slow rate of crack extension.

Typical properties for fibre fillers used in reinforcement of plastics are given in Tables 3.5 to 3.9.

Table 3.6 Cloths and mats for high toughness performance.

Chemical nature	Forms available	Typical toughness factor $\left(\dfrac{\text{Impact strength composite}}{\text{Impact strength base polymers}} \right)$
Cotton	Fabrics	10–20
Asbestos	Fabrics and mats	10–30
Nylon	Fabrics	10–20

Table 3.7 Short fibrous fillers for medium stiffness and strength performance

Chemical nature	Forms and types available	Comments
Glass fibres (E)	Chopped rovings	Chemical couplings on the surface
Asbestos fibres	Classified grades	Dispersion difficult (some water adsorption)
Mica (flakes)	—	Wetting and adhesion difficult
Glass (flakes)	—	—

Table 3.8 Short fibrous fillers for medium toughness performance

Chemical nature	Forms available	Comments
Cellulose types	Cotton flock and chopped fabrics. Paper and wood pulp. Rayon fibres and fabrics. Jute, hemp and sisal thread.	Main use in phenolics and amino plastics. High water absorption.

Table 3.9 Low reinforcement (cheapening) particulate fillers

Filler	Comments
Talc	
Silica flour ⎫ China clay ⎬	'Slightly' reinforcing fillers
Calcium carbonate ⎫ Chalk ⎬ Barytes ⎭	'Cheapening' fillers

References

1. VINCENT, P. I., *Polymer,* 12, 1971, p. 534.
2. FERGUSON, R. J. and WILLIAMS, J. G., *Polymer,* 14, 1973, p. 103.
3. EINRICH, F. R., *Applied Polymer Symposium 1,* 'High speed testing', Vol. V, Interscience, 1965.
4. BIKERMAN, J. J., *SPE Transactions,* V4, 1964, p. 267.
5. WURSTLIN, F., *Kolloid-Zeitschrift,* 33, 1957, p. 1.
6. LEUCHS, O., *Kunstoffe,* 46, 1965, p. 12.
7. SELKER, M. L., WINSPEAR, G. G. and KEMP, A. R., *Ind. Eng. Chem.,* 34, 1942, p. 157.
8. CLASH, R. F. (Jr.) and BERG, R. M., *Ind. Eng. Chem.,* 34, 1942, p. 1218.
9. NIELSON, E. L., *Mechanical Properties of Polymers,* Reinhold, 1965, p. 168.
10. FRISSEL, W. J., *Modern Plastics,* 38, May, 1961, pp. 232–8.
11. DOOLITTLE, A. K., *Journ. Poly. Sci.,* 2, No. 2, 1947, p. 121.
12. GHERSA, P., *Modern Plastics,* 36, No. 2, October, 1958, p. 135.
13. JACKSON, W. J. and CALDWELL, J. R., *J. Appl. Poly. Sci.,* 11, 1967, pp. 211–44.
14. HORSELY, R. A., British Plastics Convention, London, 11th–17th July, 1957.
15. McCULLOUGH, (REV.) R. L., *Concepts of Fibre-Resin Composites,* Dekker, 1971.
16. SHAFFER, B. W., Materials Properties of Reinforced Plastics, *SPE Transactions,* SPI, 4, 1964, p. 267.
17. CALCOTE, L. R., *The Analysis of Laminated Composite Structures,* Van Nostrand–Reinhold, 1969.
18. ASHTON, J. E., HALPIN, J. C. and PETIT, P. H., Primer on Composite Materials: Analysis, Technomic Publishing Co.
19. KELLY, A. and DAVIES, G., *Metallurgical Reviews,* 10, No. 37, 1965.
20. NIELSON, L. E. and CHEN, P. E., *Journal of Materials, J.M.L.S.A.,* 3, No. 2, 1968, pp. 352–8.
21. HALPIN, J. C., *Journ. Comp. Mat.,* 3, October, 1969, p. 732.
22. MASCIA, L. and DAVIES, I. J., Second International Conference on Asbestos Materials, Louvain University, September, 1971.
23. BODNER, A. R. and LIFSHITZ, H., Mechanics of Composite Materials, Fifth Symposium on Naval structural mechanics, May, 1967, p. 663.
24. HASHIN, Z., *Ibid,* p. 213.
25. EILSER, H., *Kolloid Z.,* 97, 1941, p. 313.

26. NICHOLAIS, L., DRIOL, E. and LANDEL, R. F., *Polymer,* **14**, 1973, pp. 21–5.
27. HARTLEN, R. C., 25th SPI Technical Conference, Reinforced Plastics/Composites Division Sect. 16B, 1970, pp. 1–9.
28. MASCIA, L. and RICHARDS, P., University of Aston in Birmingham (unpublished data).
29. GORDON, J., *The New Science of Strong Materials,* Pelican, p. 14.
30. STRELLA, S., *Applied Polymer Symposia,* 7, Interscience, 1968.
31. BUCKNALL, C. B., *British Plastics*, November, 1967, pp. 118–20.
32. BUCKNALL, C. B. and SMITH, R. R., *Polymer,* 6, 1965, p. 437.

4 SURFACE PROPERTIES MODIFIERS

The surface properties of polymers have great technological importance in three areas:

(a) friction phenomena and wear,
(b) adhesion to metals and other substrates in laminar composite products,
(c) surface conduction of electrical charges.

In all cases the properties of plastics surfaces are influenced by their roughness, polarity of chemical groupings and boundary layers of foreign matter.

It is expected, therefore, that additives may be used to alter any of these three parameters.

4.1 Additives which increase surface roughness

It is often desirable to increase the surface roughness of plastics (on a micro-scale) in order to reduce the intimacy of contact with another component or to induce scattering of the reflected light and reduce surface gloss (see Chapter 5). Such an effect can only be produced by introducing into the polymer fine, rigid, or rubbery particles. Surface irregularities are formed as a result of micro-fractures or differential relaxation and shrinkage at the melt/metal interface during processing. Surface roughness is required, for instance, in the production of packaging films in order to minimize 'blocking', i.e. the tendency of adjacent film surfaces to stick to one another under pressure.

Anti-block additives that may be used in films are, therefore, very fine fillers (dia. $\cong 0{:}01 - 0{:}01 \ \mu\mathrm{m}$), e.g. silica flour.

Rubber particles are not normally used for this purpose, although in the case of toughened thermoplastics sheets a surface mattness is often produced as a result of differential shrinkage and melt relaxation, hence, providing adequate anti-blocking characteristics.

4.2 Additives which form a boundary layer on the surface of plastics

The formation of a boundary layer on the surface of plastics resulting from the exudation of additives from the bulk is exploited industrially to reduce friction and wear and to reduce the build-up of electrostatic charges.

4.2.1 Solid lubricants

Solid lubrication is an effective method of reducing friction and wear in components such as bearings, bushings, etc. The function of solid lubricants is to produce a micro-roll-bearing action at the interface. Since the lubricant has a lower cohesive and shear yield strength than the two adjacent materials, it provides a cushioning effect between the rubbing surfaces.

External lubricants which are normally used in processing tend to be removed readily from the interface and, therefore, will only function for a short period of time.

Furthermore in many engineering applications fibre reinforced plastics are used, where the conventional processing lubricants cannot provide a sufficiently thick boundary to prevent wear owing to broken fibre debris. Consequently fairly large lubricant particles are required at the interface. The present practice is to use molybdenum disulphide, graphite and PTFE powders with both unfilled and reinforced plastics.

4.2.2 Antistatic agents

Electrification of materials results from a segregation of charges (electrons and ions) which occurs when two surfaces are parted after close initial contact. The amount of charge build-up is determined by the rate of generation and, simultaneously, the rate of charge decay.

The rate of charge generation on the surface can be decreased to some extent by reducing the intimacy of contact, whereas the rate of charge decay can be increased substantially by rendering the surface conductive by the formation of an ionic boundary layer. Consequently antistatic characteristics in plastics can be achieved by means of ionizable additives which can migrate to the surface and form conductive paths through the absorption of atmospheric moisture.

Such additives are called 'antistatic agents' and normally consist of:

(a) Nitrogen compounds: long chain amines, amides or quaternary ammonium salts, e.g. stearamido-propyldimethyl-2-hydroxyethyl ammonium nitrate.
(b) sulphonic acids and alkyl aryl sulphonates,
(c) polyhydric alcohols and derivatives,
(d) phosphoric acid derivatives, e.g. didodecyl hydrogen phosphates
(e) polyethylene glycol derivatives, e.g. hexadecyl ethers of polyethylene glycol.

Molecular weight and overall polarity considerations are important since they determine the rate of diffusion of the additive and hence their efficiency and durability. Although strongly ionic inorganic salts, e.g. LiCl, etc. would provide very effective antistatic behaviour initially, they would easily leach out of the system on subsequent ageing. Consequently a certain balance between compatibility and diffusibility must be achieved in order to obtain optimum efficiency and durability. The rate of charge decay obeys an exponential relation of the type[1]:

$$V_t = V_0\, e^{-t/RC} \quad \text{or} \quad \log \frac{V_t}{V_0} = -0.434 \frac{t}{RC} \qquad (4.1)$$

Hence the rate of charge decay is a logarithmic function of time, and the rate constant is defined as the time (τ) for the charge to reach $1/e$ of its original value. The rate constant can, therefore, be obtained from measurements of surface resistivity. Good antistatic properties are achieved when τ becomes of the order of a fraction of a second.

The relationship between antistatic rating and surface resistivity is shown in Table 4.1.

Table 4.1[1]   Relationship between surface resistivity and antistatic characteristics

$\log R$ (Ω)	τ (sec)	Antistatic protection
13	> 30 min	None
12–13	10–30 min	Poor
11–12	10–30 sec	Moderate
10–11	0.01–10.8 sec	Good
8	0.01 sec	Excellent

4.3 Additives which alter the polarity of the surface

The polarity of the surface of materials in the absence of a boundary layer can only be altered by enriching the regions of the material near the surface with additives whose polarity is substantially different from that of the basic material.

This situation is achieved if the additive is sufficiently compatible with the polymer but has a low tendency to crystallize out on the surface of the polymer, or to dissolve in the adsorbed water. In other words, it is desirable that such additives be substantially different from conventional external lubricants or antistatic agents.

4.3.1 Adhesion promoters

Bonding of plastics materials to inorganic or other substrates is an important aspect of plastics technology. It is often desirable for instance to make sandwich films, polymer/paper laminates, polymer/metal laminates, heat sealing polymeric films, etc. It is to be expected, therefore, that the adhesion of plastics to other substrates depends mainly on the types of chemical bonds present at the interface.

In practice, other factors must be considered, namely: degree of wetting or intimacy of contact of the two surfaces and adsorbed foreign matter forming interlayers between the two surfaces considered.

Adhesion promoters will, therefore, have the function of reducing the surface tension between the polymer and the substrate and/or increasing the magnitude of the forces or bonds across the interface. This can be easily shown from considerations of the classical theory of wetting.

An element of highly deformable material in contact with a rigid substrate will form an angle θ, as shown in Fig. 4.1. Resolving the

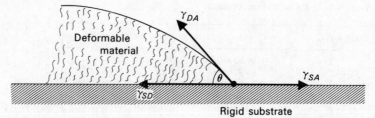

Fig. 4.1 Idealized behaviour of wetting of rigid substrate by a highly deformable body.

forces (surface tensions γ) acting in the interface plane we obtain

$$\gamma_{SD} + \gamma_{DA} \cos \theta = \gamma_{SA} \tag{4.2}$$

hence

$$\cos \theta = \frac{\gamma_{SA} - \gamma_{SD}}{\gamma_{DA}} \tag{4.3}$$

where S = substrate

D = deformable body

A = a third phase

Therefore the coverage of the substrate by the deformable body (e.g. polymer/adhesion promoter interlayers) is highest when θ tends to zero. Consequently in order to reduce θ the term γ_{SD} (the surface tension between polymer interfacial layer and the substrate) must decrease.

If the substrate is highly polar (e.g. of inorganic or cellulosic nature) and the polymer is non-polar (e.g. polyolefins) wetting and interfacial bonding will increase by incorporating of compatible polar additives, containing carboxylic acid, amine, amides or urethane groups. To increase wetting it would be desirable to have regions at the interface which are highly plasticized (easily deformable), and to obtain maximum bonding it is essential that the polar groups of the additive become aligned with those on the surface of the substrate. Both suggest that lower molecular weight substances are preferred since they will have greater diffusibility and will more readily migrate to the interface. These will also exhibit a lower conformational energy which allows them to become aligned on the surface of the substrate. If multi-molecular layers are formed at the interface, however, it is likely that the shear strength of these is lower than that of either of the two adjacent materials and therefore a low bond strength will result. Hence the concentration and the relative compatibility of such additives is very critical. With thermoplastics, since they are normally applied on to substrates at high temperatures, it is desirable to choose adhesion promoters which act primarily as high temperature plasticizers. Stearic acid, for instance is a notably good adhesion promoter for polyethylene towards metal substrates.

Commercial adhesion promoters

For many purposes the substances normally used to aid adhesion are similar to those described to increase the interfacial bonding in composites (Chapter 3, p. 90). There are others, however, which are becoming increasingly important commercially, namely:

(a) Phosphorous containing compounds.

These have been suggested especially to aid adhesion of polymers on metal substrates because of their chelating action towards metal ions.

(b) Acids.

Several of these have been used including stearic acid, salicyclic acid and p.-chlorophenyl substituted fatty acids[2].

(c) Amines.

Of these the most widely used are the piperidine derivatives[1] because the polar amine groups are not sterically hindered and can therefore provide good alignment on the surface of substrates.

(d) Titanates.

These have also been reported to aid adhesion to glasses and metals. It has been suggested that their mechanism may be associated with co-ordination reactions with the substrate[2] i.e.

$$
\begin{array}{ccccc}
\text{OR} & & \text{OR} & & \text{OR} \\
| & & | & & | \\
-\text{Ti}-\text{O}-\text{Ti}-\text{O}-\text{Ti} \\
| & & | & & | \\
\text{O} & & \text{O} & & \text{O}
\end{array}
$$

4.4 Evaluation of surface modifiers

Modification of the surface properties in effect means altering the surface free energy, irrespective of whether such changes involve increasing the surface roughness or its polarity and hydrophilic characteristics.

In each case the decrease in contact angle which a water droplet will form on the surface can be used to monitor any such changes produced by the introduction of additives.

Measurement of the contact angle can be easily carried out by means of a travelling microscope fitted with a protractor at the eye piece. Using Wenzel's[3] definition of surface roughness factor r and its relationship to contact angle,

$$r = \frac{\text{true surface area}}{\text{apparent plane area}}$$

and

$$\cos \theta_{(r)} = r \cos \theta \qquad (4.4)$$

where θ = contact angle on smooth surface
   $\theta_{(r)}$ = contact angle on rough surface

the method could also be used as a screening procedure for the evaluation of anti-block additives etc.

On the other hand, surface roughness will also reduce the light reflection properties of the material, hence it can be evaluated by monitoring the loss of intensity of light reflected from an incident light source impinged at $45°$ on the surface.

The efficiency of a particular surface modifier ulimately will be assessed in terms of its 'true' function. Adhesion characteristics will be assessed by measuring peel strength, antistatic properties by measuring surface resistivity, and interfacial friction and wear resistance tested by the usual sliding members and abrasion methods.

References

1. *Encyclopaedia of Polymer Science and Technology*, Vol. 2, Interscience, 1965, p. 213.
2. SKEIST, I., *Reviews in Polymer Technology*, Vol. 1, Dekker, 1972, p. 19.
3. KAELBLE, D. H., *Treatise on Adhesion and Adhesives* (R. L. Patrick, Ed.), Vol. 1, Edward Arnold, 1967, p. 175.

5 OPTICAL PROPERTIES MODIFIERS

The whole subject of the optics of materials is rather specialized and complex, and a detailed account would be beyond the scope of this book. Here we are concerned with some of the basic principles which are of importance in the technology of plastics from the point of view of improving their visual appearance.

The aesthetic value of objects, in fact, is invariably judged by their response to visible light (wave length range from 400 to 700 nm) and by their geometrical shape (i.e. the stimulus produced at the retina of the eye, which is then interpreted by the brain).

The optical properties which determine the aesthetic quality of plastics products can be described in terms of their respective ability to transmit light, to exhibit colour and to reflect incident light.

5.1 Light transmission of plastics

When all the incident white light is transmitted through a material and remains essentially unchanged, the material is said to be 'transparent'. Transmission depends on the amount of reflection at the surface, which in turn depends on the refractive index n_D of the material and on the angle of incidence.

The amount of light reflected is given by the Fresnel equation[1]:

$$R = \frac{1}{2}\left[\frac{\sin^2(i-r)}{\sin^2(i+r)} + \frac{\tan^2(i-r)}{\tan^2(i+r)}\right] I_0 \qquad (5.1)$$

where I_0 = intensity of incident light

$\quad i$ = angle of incidence

$\quad r$ = angle of refraction

$\quad n_D$ = refractive index = $\dfrac{\sin i}{\sin r}$

For incident light perpendicular to the surface ($i = 0$) a simplified equation for light transmission can be obtained:

$$\% \text{ transmission} = \left[1 - \frac{(n_D - 1)^2}{(n_D^2 + 1)} \right] \times 100 \qquad (5.2)$$

The percentage of light transmitted depends also on the amount of light scattered internally and at the surface of the material. Scattering of light is caused by the random reflection on the surface irregularities and at the interface of small particles or voids embedded in the material, owing to the changes in directions of light as it passes through media of different refractive index.

The refractive index of the two phases in polymer blends (e.g. acrylics/PVC blends) may change at different rates when the temperature is increased, and may result in further loss of optical clarity (Fig. 5.1). At the exit of a die a PVC blend extrudate is often "milky" but it becomes transparent on subsequent cooling.

In general, light scattering depends on the size of scattering centres and is greatest when the size of the scattering centre is of the same order of magnitude as the wave length of incident light[2]. Scattering reaches a maximum when the size of particles is about half the wave length of light (i.e. 200–400 nm) and reduces drastically when the size of scattering centres becomes appreciably smaller than the wave length of light (Fig. 5.2).

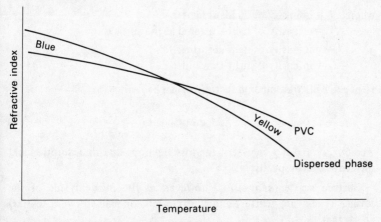

Fig. 5.1 Differential temperature coefficient for dispersed phase and matrix. (After Ryan, C. F., Poly-blends and Composites, *Applied Polymer Symposia,* **15**, p. 179.)

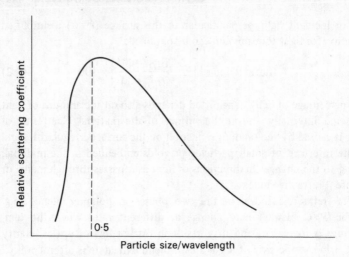

Fig. 5.2 Scattering as a function of particle size for TiO_2. (After Mitton and White, *Official Digest*, **30**, 1958, p. 1268.)

With homogeneous materials, or when the internal scattering centres are smaller than the wave length of light, the transmission properties of the material can be related to the incident light and to the absorbed and scattered light by a simple expression[3]:

$$I_i = I_{sc} + I_{abs} + I_{tr} \tag{5.3}$$

where I_i = intensity of incident light
I_{sc} = intensity of light scattered at the surface
I_{abs} = intensity of light absorbed
I_{tr} = intensity of light transmitted.

Hence a 'light transmission factor', T, can be defined as:

$$T = \frac{I_{tr}}{I_i} = 1 - F_{abs} - F_{sc} \tag{5.4}$$

where F_{abs} and F_{sc} represent the loss fractions due to absorption and surface scatter respectively.

Surface scatter can occur, however, at the incident side of the surface (backward scatter) and the transmission side (forward scatter) i.e. so that

$$I_{sc} = I_{sc}{}^b + I_{sc}{}^f \tag{5.5}$$

hence

$$[I_{tr}]_{\text{tot}} = I_{tr} + I_{sc}^f \qquad (5.6)$$

and the forward scatter fraction fraction X_{sc}^f is defined as:

$$X_{sc}^f = \frac{I_{sc}^f}{[I_{tr}]_{\text{tot}}} \qquad (5.7)$$

The loss of contrast which occurs when an object is viewed through a plastics sheet or film is due to forward scatter if the illuminating source is on the same side as the object, and the effect is called 'external haze'.

The loss of contrast which takes place when objects are viewed through plastics sheets from the same side as the illuminating source is due to the combined effect of forward and backward scatter and is normally called 'milkiness'.

In absence of appreciable internal scatter, the size of the scattering centres on the surface (i.e. surface roughness) can also have a considerable effect on the optical resolving power of the plastics material.

Large surface irregularities, in fact, impair the resolution of fine details of objects when viewed at some distance away from the plastics material. This is due to forward scattering occurring at very small angles (i.e. between $1°$ and $1·5°$) from the surface and is normally known in the packaging industry as the 'see-through-clarity'. External haze and milkiness, on the other hand, are effected by fine surface irregularities which produce wide angle scatter (i.e. $> 2·5°$) from the surface of the material.

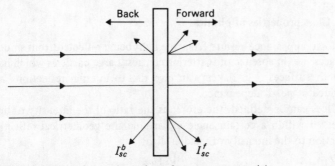

Fig. 5.3 Light scattering on the surface of transparent materials.
(After Birley A. W. and Ross G. The Institute of Physics Conference at Buxton, 5th–6th October 1972.)

5.2 The colour of plastics

Colour in a material is developed as a result of visible light being absorbed. Absorption of light occurs through dissipation of electromagnetic energy by the electronic configuration of the material, (i.e. electrons acquire energy in excess of that in their 'ground' state). When all the light is absorbed by the material it appears 'black'. If only part of the light is absorbed (i.e. only light of certain wave lengths) and the amount of light scattered is small the material becomes 'coloured transparent' and the colour developed corresponds to the particular wave lengths of light transmitted.

If the amount of light which has not been absorbed is internally scattered then the material becomes 'coloured opaque' and the colour corresponds to that of the wave lengths not absorbed, although a more complex response can often take place and a different colour may result. The generally accepted ranges of wave length corresponding to the common colours of the visible light spectrum is shown in Table 5.1.

Table 5.1 Relationship between spectrum colours and wavelengths

Spectrum colour	Wavelength (nm)	Spectrum colour	Wavelength (nm)
Red	750–610	Green	555–510
Orange	610–590	Blue–Green (Cyan)	510–480
Yellow	590–575	Blue	480–450
Yellow–green	575–555	Violet	450–400

5.3 Gloss properties of plastics

A glossy appearance results from the light being reflected from smooth surfaces, i.e. in absence of scattering centres. Large particles which have smooth surfaces (e.g. flakes) can give rise to internal reflections and produce a 'pearl' appearance.

Gloss can be defined, therefore, as the ratio of the intensity of light reflected within a certain angle ω_r around the geometrical reflection direction to the intensity of incident light,

$$G = \frac{[I_r]\,\omega_r}{I_i} \tag{5.8}$$

For practical measurements the angle ω_r normally used is $45°$.

5.4 Additives which alter the light transmission characteristics

Plastics materials have a refractive index between 1.45 and 1.70 and, in the absence of light scattering, the transmission is in the region of 80 to 90% of the incident light. All amorphous polymers have 'good' light transmission characteristics. Crystalline polymers, on the other hand, tend to develop internal haze due to the difference in refractive index between the spherulites (larger than the wave length of light) and amorphous regions. Since the crystal regions and the amorphous domains have the same chemical constitution, the only factor responsible for the difference in refractive index is the density. The light transmission properties are, therefore, mainly a function of the difference in density (see Table 5.2).

Table 5.2

Material	Crystalline regions density	Amorphous regions density	Optical transmission
Polyethylene	1.01	0.84	Opaque/translucent
Polypropylene	0.94	0.836	Translucent
Poly,3,Me-butene-1	0.90	0.836	Translucent/transparent
Poly,4,Me-pentene 1	0.83	0.83	Transparent

In the case of crystalline polymers having spherulitic entities greater than the wave length of light, there is also a considerable contribution from internal scatter occurring on the surface of spherulites.

The natural colour of polymers varies, therefore, from 'water clear' for amorphous materials to 'white opaque' in the case of highly crystalline polymers (e.g. high density polythene, acetals, etc.), and 'yellow amber' for those polymers based on phenolic compounds, owing to quinoid structures being formed during polymerization and on ageing. A yellow tint can also develop as a result of conjugated double bonds and ketonic carbonyl groups being formed during processing and ageing (see Chapter 6, p. 133).

To aid transmission of white light, therefore, the following possibilities are open:

(a) reduce the extent of crystallization,
(b) promote the formation of the crystal modifications of lowest

118

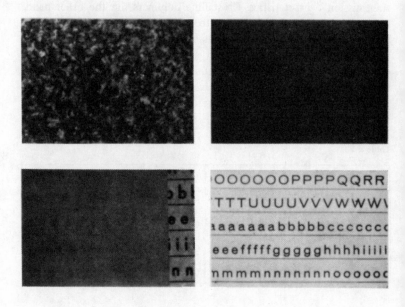

Plate 5 Relationship of spherulitic size (100 x magnification) and clarity of nucleated polypropylene (right).
(After Beck, H. N. and Ledbetter, H. D., *J. Appl. Poly. Sci.,* 9, 1965, p. 2138)

density in the case of polymorphic polymers (e.g. stereoregular polymers),
(c) reduce the size of spherulites to values below the wave length of visible light,
(d) absorb the light in the yellow-amber region of the spectrum (575–625 nm) and re-emit it in the non-visible region.

At present only a limited number of additives are capable of aiding transmission of white light in plastics, namely 'nucleating agents', which function by promoting the formation of a large number of nuclei and reducing the average size of spherulites in crystalline polymers (see Plate 5), and 'optical brighteners' which mask some of the yellow discolourations produced during processing. Optical brighteners are fluorescent organic substances which absorb ultra-violet radiation in the far end of the visible spectrum ($\cong 300–400$ nm) and re-emit it in the lower end of the spectrum ($\cong 450–550$ nm).

Examples of nucleating agents and optical brighteners are shown in Table 5.3.

Table 5.3

Optical brighteners	Nucleating agents
1 Benzosulphonic and sulphonamides derivatives of 4-naphtholtriazolyl stilbene	1 Na, K, Li Benzoate
2 Vinylene bisbenzoxazoles	2 Finely divided inorganic powder (particle size below 40 nm) e.g. clays, silica flour, etc.
3 4-alkyl-7-dialkyl amino coumarins, and other derivatives	

To impart colour to plastics without appreciably affecting their transparency, additives must be used which will absorb unwanted wave lengths of the visible spectrum and allow the others through. It is necessary that such additives are compatible with the polymer or can be so finely dispersed that they will not act as scattering centres.

Such substances are called 'dyes' and are normally organic substances whose refractive index is not too different from that of plastics so that the actual size of particles, when they are not soluble in the polymer, is not so critical for light transmission[4].

There are obvious economic and technical difficulties in producing a range of dyes wide enough to obtain all the possible shades of colours, hence it is very common to use mixtures.

Colour mixture experiments with light of narrow wave lengths have shown that the common colours can be produced (matched) by a mixture of three colours called 'primaries', which are red, blue and green. Each primary colour takes up one-third of the spectrum[5] (Plate 6) and is chosen so that a combination of all three (in equal amounts) will match the total spectrum of white light, and no combination of any two will produce the third primary colour.

Combination of primaries to produce a different colour is seen as an additive process and is, therefore, called 'additive colour mixing'. The colours produced by equal pair combination of primaries are:

$$Red + Blue \quad = Magenta$$
$$Red + Green = Yellow$$
$$Blue + Green = Cyan$$

and are known as 'secondaries'.

Additive colour mixing can be represented in the form of a colour triangle[6], with primary colours at the corners and the spectrum colours on a curve located by the relative amounts of the three primaries required to match them (Fig. 5.4). White light is at the centre of the triangle because it is matched by equal amounts of all three primaries, whereas any other colour C will be matched by amounts a, b and c of blue, red and green respectively. Intermediate parts of the spectrum cannot be matched by all three primaries because the curve for the spectrum lies outside the colour triangle.

A second method, called 'subtractive colour mixing' is also used for colouring purposes. It relies on the fact that a combination of secondary colours can produce other colours (Plate 6) and when viewed subtractively a primary colour will result. For instance, magenta and yellow will produce red, since red, and only red, can be transmitted by both.

By using, therefore, the three secondaries either alone or in pairs, six colours are produced (Tables 5.4 and 5.5 and Fig. 5.5). By using them in varying relative amounts a complete colour circle of 'hues' can be obtained.

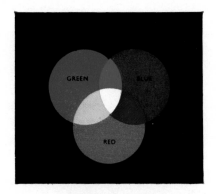

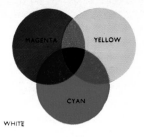

A Primary lights superimposed additively
B Secondary colours superimposed subtractively, as in printing, etc.

Mixture of (A) primary and
(B) secondary colours

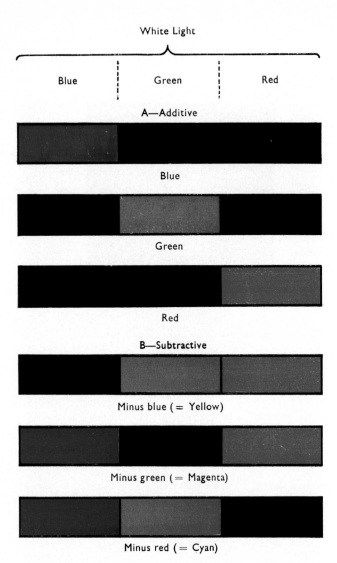

White Light

Blue | Green | Red

A—Additive

Blue

Green

Red

B—Subtractive

Minus blue (= Yellow)

Minus green (= Magenta)

Minus red (= Cyan)

Diagrammatic illustration of the transmissions of (A) primary and (B) secondary colour filters. When superimposed in pairs, B have common transmission bands, whereas A do not

Note: These are idealised conditions; real filters do not have such sharply defined absorption boundaries.

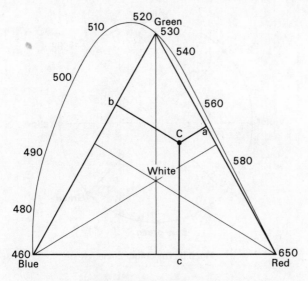

Fig. 5.4 The colour triangle.

Table 5.4 Secondary colours[5]

Colour seen	Light transmitted	Light absorbed
Magenta	Blue and red	Green
Yellow	Red and green	Blue
Cyan	Green and blue	Red

Table 5.5 Mixture of secondary colours[5]

Mixture	Light transmitted (= colour seen)	Light absorbed
Magenta + Yellow	Red	Green + blue
Yellow + cyan	Green	Blue + red
Cyan + magenta	Blue	Red + green
Magenta + yellow + cyan	None (= black)	Red + green + blue

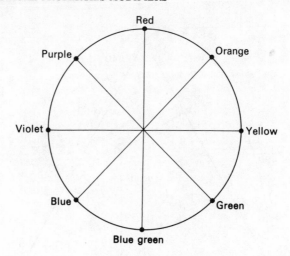

Fig. 5.5 The colour circle.

Furthermore we can take into account the 'value' of colour and the 'hue' by constructing a 3-dimensional diagram, Fig. 5.5. By making one of these colours duller, i.e. by addition of a suitable proportion of black, a 'tertiary' colour will result. Since the three secondaries together make black, dullness is produced by adding the right amount of one or two extra secondaries to make up three in all (Table 5.6).

Table 5.6 Tertiary colours[5]

Original colour seen	Produced by using	Made duller by adding a little	Final colour seen
Red	Magenta + yellow	Cyan	Maroon
Yellow	Yellow	Cyan + magenta	Brown
Green	Yellow + cyan	Magenta	Olive
Blue	Cyan + magenta	Yellow	Navy

On the other hand, paler depths of any colour are obtained by adding a proportion of white or by using smaller quantities but maintaining the ratios of the component secondaries constant. In selecting dyes one would, therefore, be guided by the principles illustrated in tables 5.4, 5.5 and 5.6.

Dyes used in plastics are normally divided into two categories 'spirit soluble' (i.e. soluble in alcohols, ketones, esters, and ethers) and 'oil soluble' (i.e. soluble in aromatic compounds, containing amino, nitro

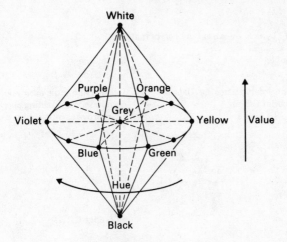

Fig. 5.6 Munsell colour system.

groups, etc.)[7]. Selection of a dye for a given polymer system is made according to the mutual solubility of the dye and the polymer in one of the solvents mentioned. Typical dyes used in plastics are given in Table 5.7

5.5 Additives which impart opacity and colour

As already discussed opacity can only be obtained through reflection, scattering and absorption of light. Scattering and reflection play an important role when only partial absorption of the light takes place, i.e. for any colour other than black. Consequently a main requirement for additives used to impart colour and opacity is that their particle size should be above the wavelength of light and their refractive index much higher than that of plastics. Such substances are called 'pigments' and can be either inorganic or organic in nature. The number of inorganic pigments available is rather limited and consequently the majority of pigments are obtained from dyes which are insolubilized by means of binders (thermosetting resins or gums) and converted to particles of the desired size. Because of the interaction of scattering and absorption, prediction of colours from mixtures of pigments can be much more complex than the simple additive and subtractive mixing relationships. Examples of pigments commonly used in plastics are shown in Table 5.8.

Table 5.7 Typical dyes used in plastics materials[7]

Spirit soluble	Oil soluble
1 Salts of organic bases e.g. Zapon fast scarlet GG	Various non-ionic aromatic compounds containing alkyl groups to improve compatibility e.g.

$[(C_6H_{11})_2\overset{\oplus}{N}H_2]_2$

oil orange C.1.24

2 Metal complexes of azo dyes e.g. Zapon fast yellow GR

Oil yellow

3 Derivatives of triphenyl methane e.g. spirit blue C.1.689

Sudan violet R

Sudan blue GL

Table 5.8 Common pigments used in plastics products

Inorganic pigments	Comments
1 Titanium dioxide (rutile, anatase)	High covering power due to its very high refractive index. Widely used for whiting pigmentation.
2 Cadmiun sulphides, sulpho-selenides	Used for yellow-orange shades. May accelerate polymer degradation.
3 Lead chromate, molybdate	Used for yellow-orange shades.
4 Chromium oxide	Green pigments. Good light fastness.
5 Ultramarine blues (sodium aluminium, polysulphide silicates)	Reddish blue violet shades.

Organic pigments	
1 Benzidene and derivatives	Used for yellow-orange shades. Good heat stability.
2 Copper phthalocyanine	Used for blue shades. Good heat stability.
3 Chlorinated copper phthalocyanine	Used for green-blue shades. Good heat stability.
4 Aniline black	Used for black-bluish green shades.

Pigmented lakes	
These are water soluble acid dyes converted into insoluble pigments by precipitation on to an inorganic base, such as alumina.	Relatively low tinctorial strength and light stability, generally lower than organic pigments[9].

Lamellar pigments	Comments
1 Basic lead carbonate	Provide a pearl appearance owing to the high reflectivity of their surface and their high refractive index ($\cong 2$).
2 Lead hydrogen phosphate and arsenate	
3 Bismuth oxychloride	
4 Aluminium flakes	
5 Copper alloys (largely Cu-Zn) flakes	Gold-bronze colours.
6 Copper flakes	Impair heat and light stability.

5.6 Colour measurement and evaluation of dyes and pigments

Measurement of colour of materials is normally made by means of colorimeters and spectrophotometers.

Colourimetric methods consist of matching the colour quality of the sample by an additive mixture of three primary colour lights, red (R),

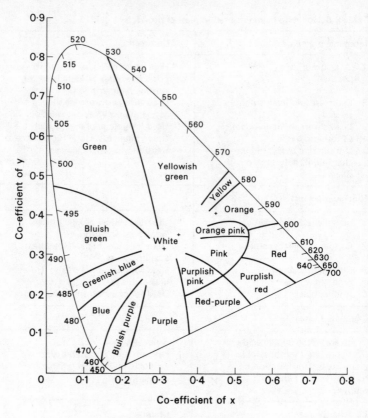

Fig. 5.7 The C.I.E. chromaticity diagram.

green (G) and blue (B) and specifying the dullness or brightness, which is a measurement of the reflectance (Y).

Colour quality is represented by the relative values of R, G and B and the quantities of R, G and B required for the match, i.e. by the three derived values x, y and z defined as:

$$x = \frac{R}{R+G+B}; \quad y = \frac{G}{R+G+B} \quad \text{and} \quad z = \frac{B}{R+G+B}$$

Now, since $x + y + z = 1$, only two of these need to be specified (the third can be determined by difference). Normally x and y are plotted on a diagram similar to Fig. 5.5, but using rectangular axes. Such a plot is called the C.I.E. (Commission Internationale de l'Eclairage) chromaticity diagram (Fig. 5.7). The dullness or brightness is expressed in terms of a reflectance or luminance factor (% reflected light from a standard magnesium oxide screen). (An example of colour specification is BS 381 C:538. Post Office red $x = 0.656, y = 0.318$ and $Y = 8.5$.)

With spectrophotometers a continuous record of transmittance or reflectance is obtained over the visible wavelength range and the relative proportions of each colour absorbed or reflected determined (see Table 5.1).

For evaluation and formulation purposes it is quite common practice in industry to use trial-and-error methods. Such a practice, however, may lead to usage of a larger number and greater amounts of expensive dyes or pigments than would be used if a more rational approach was adopted.

For dyeing purposes the relative levels of various dyes can be predicted from spectrographic measurements of the absorbance or optical density ($d = \log I_0/I_t$, where I_0 = intensity of incident light and I_t = intensity of transmitted light) of the colour to be matched. For three colour mixtures (the maximum number normally required) the optical density at any given wavelength is related to the concentration of each individual dye by:

$$d = \alpha_A C_A + \alpha_B C_B + \alpha_C C_C$$

where

(i) α_A, α_B and α_C are the absorptivities for dyes A, B and C respectively, defined by Beer's law ($\log I_0/I_T = \alpha C x$.)

(ii) C_A, C_B and C_C are the respective molar concentrations.

(iii) x is the thickness of the sample.

For pigmentation purposes, on the other hand, it is necessary to take into account the scatter coefficient (δ) in addition to the absorption coefficient (a). According to Duncan[8] both absorptivity and scatter coefficients of pigment mixtures are additive and the reflectance (R) at any given wavelength can be calculated:

$$\phi = \frac{C_A \delta_A + C_B \delta_B + C_C \delta_C + \cdots}{C_A \alpha_A + C_B \alpha_B + C_C \alpha_C + \cdots}$$

where $\phi = 2R/(1 - R^2)$

In either case, however, calculations alone will not provide an exact estimate of the most suitable types and levels of dyes and pigments and a final tinting will invariably be required to make a precise match[6]. This arises from small unavoidable colour variations due to differences in tinting strengths, dye or pigment dispersion, internal scatter due to polymer spherulites, insoluble additive and impurities such as lubricants etc.

In an evaluation programme concerning the use of dyes and pigments, however, it is not a matter of simply deciding what combination and levels of additions are required to produce the right colour. In fact, although colour matching in itself is by no means a simple task, the major problems that a polymer technologist may have to overcome are associated with side effects which may impair the overall performance of the material. Factors such as compatibility, decomposition during compounding and processing, chemical interactions with other components, (e.g. stabilizers) blooming and effects on ageing characteristics, may be more decisive in the final selection than those based on colour considerations alone.

Metal salts and chelates in a pigment mixture can accelerate degradation of polymers such as acetals, polyolefins and the products of the decomposition of the polymer, e.g. PVC, cellulose acetate, etc., may cause a breakdown in the chemical structure of the colouring matter.

References

1. BAER, E., *Engineering Design for Plastics,* SPE Polymer Science and Engineering Series, Reinhold, 1964, p. 603.
2. MITTON, P. B and WHITE, L. S., *Official Digest,* **30,** 1958, pp. 1259–1276.
3. ROSS, G., and BIRLEY, A. W., The Institute of Physics Conference at Buxton, 5th–6th October, 1972.
4. BILLMEYER, F. W., and SALTSMAN, M., *Principles of Colour Technology,* Wiley Interscience, 1967, p. 94.
5. GILES, C. H., *A Laboratory Course in Dyeing,* The Society of Dyers and Colourists, 1971 Edition, p. 18.
6. The Research Association of British Paint, Colour and Varnish Manufacturers, *Colour in Surface Coatings,* 1956, pp. 45–67.
7. WARDLE, S. D., Manchester Polytechnic, Personal communication.
8. DUNCAN, D. R., *J. Oil Col. Chem. Assoc.* 1949, **32,** p. 296.
9. RENFREW, A., and MORGAN, P. *Polythene,* Iliffe Books Ltd., 1957, p. 433.

6 ANTI-AGEING ADDITIVES

Ageing may be defined as the process of deterioration of materials resulting from the combined effects of atmospheric radiation, temperature, oxygen, water, micro-organisms, and other atmospheric agents (e.g. gases).

Although it could be argued that deterioration and failure of materials can result from the action of other external agents, such as mechanical stresses, etc., it is customary to use the term ageing to indicate that a chemical modification in the structure of the material has occurred. The manner in which each of the above factors cause deterioration of plastics is determined by the chemical nature of the polymer and other basic constituents. The final effect on properties, on the other hand, is invariably the same, namely development of colour and/or reduction of optical clarity, embrittlement and increased dielectric losses.

6.1 Effects of natural radiation on ageing of plastics

Atmospheric radiation comprises ultra-violet (UV), visible and infra-red (IR) light. Of these, ultra-violet radiation is the most harmful source of radiation. The wavelength of light available on the surface of the earth which has the most harmful effect on polymers is in the 290–400 nm range of the UV spectrum. Although only about 5% of the total sunlight falling on the earth's surface is within this wave length range, most polymers contain chemical groupings or additives which can absorb radiation in this range, and therefore degradation reactions will take place. The UV light energy is normally greater than the level required to cause breaking of chemical bonds in the polymer chains (Table 6.1).

As in the case of thermal degradation discussed in Chapter 2, polymer molecules become activated or excited (i.e. acquire excess energy) as a result of absorption of UV light. This excess of energy causes breakdown of the weakest chemical bonds in the chain and 'active' free radicals are formed which initiate the degradation process.

The free radicals can produce unsaturation, chain-scission, or cross-linking of polymer molecules. In the presence of oxygen, other chemical groupings such as carbonyl, carboxyl and peroxides may be formed. Many other reactions can also occur, as for instance hydrolysis of ester or amide groups which result in chain-scission. These reactions are often accelerated by other atmospheric agents such as water, CO_2, SO_2, NO_2, etc. Examples of degradation reactions induced by radiation are shown below:

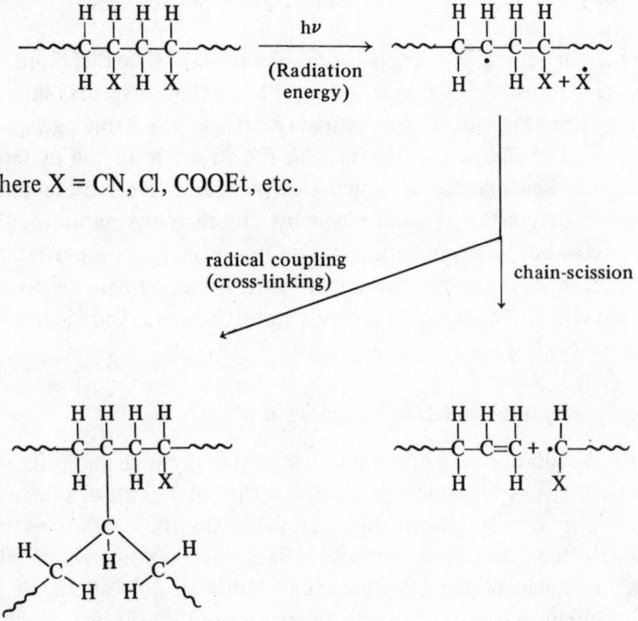

where X = CN, Cl, COOEt, etc.

Also

In presence of oxygen the degradation reactions proceed in a manner similar to thermal oxidation

Table 6.1 Energy content of light in the solar ultra-violet region and bond strength

Wavelength (nm)	Radiation energy level (kJ/einstein)	Bond type	Bond strength (kJ/mol)
290	418	C–H	355–418
300	397	C–H	314–335
		C–O	314–335
350	340	C–Cl	293–360
400	297	C–N	250–272

Note:
(a) 1 einstein = 1 'mole' photons or 6×10^{23} photons.
(b) The presence of other substituents can affect the bond strength; i.e. the C–N bond energy in amides ($-\overset{\underset{\parallel}{O}}{C}-NH-$) is reduced to about 200 kJ/mol.

(Reproduced from *Encyclopaedia of Polymer Science and Technology*, Vol. 14, pp. 127–132.)

6.2 UV protecting agents

From the preceding discussions, it can easily be deduced that to protect plastics against UV light by the means of additives, it would be necessary to reduce the energy level of the harmful radiations. The additive must, therefore, be able to absorb UV light (without undergoing decomposition) more readily than the polymer, or the additive must interact and deactivate free radicals immediately they are formed.

Those additives capable of absorbing UV radiation are called 'UV absorbers' or 'screening agents', whereas those which can offer a stabilizing effect by interaction with radicals are called 'excited state quenchers'.

6.2.1 UV absorbers or screening agents

(a) *Carbon black and pigments*

Pigments capable of absorbing strongly in the UV region will provide good protection for plastics. Carbon black can absorb over the entire range of ultra-violet and visible radiation and transforms the absorbed energy into less harmful infra-red radiations. There is also evidence however that carbon black possesses the ability to trap free radicals[1].

Substances which can reflect or scatter ultra-violet and visible radiation will also offer some protection, but are less effective than absorbers (Fig. 6.1).

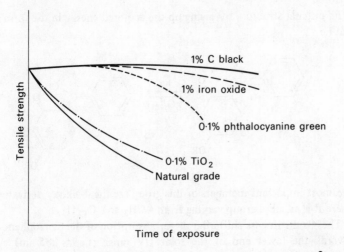

Fig. 6.1 Effect of pigments on weathering resistance of polythene[2].

Pigments however, may contain metal ions which accelerate radical formation (e.g. cobalt blue, TiO_2, etc.). Consequently these are effective only when the particles are coated so that solubilization and diffusion of metal ions is prevented. To effectively scatter UV radiation they must be used at fairly high concentrations.

Particle size of pigments and their degree of dispersion in plastics components is obviously an important consideration, but there are still many controversial arguments as to the range of size and structure of particles which yield optimum performance.

(b) *Compatible UV light absorbers*

The limitations of carbon blacks and pigments are self-evident when transparent or translucent materials are required. In these cases compatible additives must be used.

Compatible UV absorbers can be subdivided into four major classes.

Class 1 absorbers   These are derivatives of 2-hydroxy benzophenone. The UV absorption and photostabilization characteristics of these compounds are due to their highly conjugated structure and to their intramolecular hydrogen-bonding respectively. The transfer of UV energy to less dangerous low-energy quanta is accomplished thanks to its ability to be rearranged into a quinoid structure[3], which reverts back to its original structure by giving up the acquired energy in the form of heat i.e.

$$\xrightarrow[(-\text{heat})]{h\mu}$$

The most important members of this group are the 4-alkoxy derivatives where R is an alkyl group varying from $-CH_3$ to $-C_{12}H_{25}$.

The 4-substituents shifts the major absorption of the benzophenone towards the lower end of the solar UV range (i.e. $\cong 285$ nm) and increases the absorptivity over the critical range (290–400 nm). Further substitutions in the ring can accordingly cause even greater shifts in the absorption bands, and can increase UV absorption in the critical region.

Class 2 absorbers These are hydroxyphenylbenzotriazoles:

where[4] X is H or Cl
R is H or alkyl
R' is alkyl

The function of chlorine is to shift the absorption to longer wavelengths, while the function of R and R' is to increase the compatibility of the additive with thermoplastics and to reduce the volatility of the additive. There is no clear explanation regarding the mechanism of the action of benzotriazole compounds. However the formation of intramolecular hydrogen bonding and of Zwitter ions, having a quinoid structure similar to the 2-hydroxybenzophenone compounds, suggests that there may also be a similarity in their mechanism for the UV light absorption:

Class 3 absorbers These comprise esters of benzoin acids, salicyclic acid, terephthalic and isophthalic acids with resorcinol and phenols. The most important members of this class are:

(a) Resorcinol monobenzoate.   (b) Phenyl salicylate and derivatives.

(c) Di-aryl terephthalates or isophthalates.

These compounds have relatively low absorptivity in the solar UV region (Fig. 6.2) but after exposure can undergo rearrangements to 2-hydroxy benzophenones which will then absorb strongly in the harmful wavelength range.

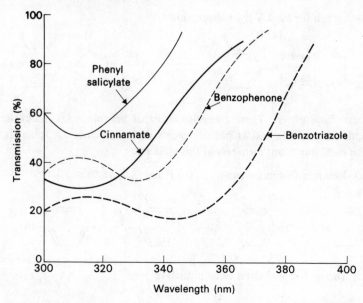

Fig. 6.2 Ultraviolet transmission characteristics of the major four classes of compatible UV absorbers[5].

The conversion however is not complete and the quinoid by-products formed may impart discolourations.

Class 4 absorbers   These additives are derivatives of Cinnamic acid:

These are also relatively ineffective UV absorbers since their maximum absorption lies in the 310–320 nm wavelength region (Fig. 6.2). Unlike the other classes of absorbers, their relative effectiveness seems to depend on the nature of the polymer in which they are used. Because of their zero-absorptivity in the visible regions, they are sometimes used when optical clarity is of utmost importance.

6.2.2 Excited-state quenchers

These additives interact with photo-excited polymer molecules and deactivate them by dissipating the excess energy as IR radiations. So far only nickel compounds have found industrial application as quenchers[4] e.g.

Briggs and McKellar[6] have found that the effectiveness of a series of experimental nickel complexes as UV stabilizers for polypropylene is similar to their quenching ability for anthracene, i.e. UV light is absorbed by the highly conjugated structure and dissipated as harmless IR radiation by a resonance stabilization mechanism[7].

These compounds however, also absorb strongly in the solar UV and some in fact (e.g. nickel di-butyldithiocarbamate) may absorb even more strongly than the benzophenones or benzotriazole compounds[8].

6.3 Commercial UV stabilizers

The use of UV radiation absorbers was introduced as early as 1945 by Meyer and Gearhart[9] to increase the service life of cellulose acetate. The first benzotriazole compound, Tinuvin P was introduced in 1956, while one year later the effectiveness of the first nickel-chelator as a UV stabilizer was discovered.

The most important types of UV stabilizers available commercially are shown in Table 6.2, of which the most important are the 2-hydroxy benzophenone derivatives.

6.4 Combined effect of oxygen and radiation

At atmospheric temperatures the reaction rates of oxygen with polymers are normally very low. In the presence of UV light however, the rate of oxidation increases considerably.

Absorption of light appears to occur in the range of wave lengths corresponding to the absorption of carbonyl and hydroperoxide groups. This indicates that the chain reactions of photo-oxidation begin with the absorption of light by carbonyl and peroxide groups already present in the polymer chains (formed by thermal oxidation during drying of the polymer after polymerization and during subsequent processing).

Table 6.2 Commercial Ultraviolet stabilizers[10]

Chemical name or description	Trade name and/or trademark
2-Hydroxybenzophenones	
2,4-dihydrobenzophenone	Advastab 48
	Rylex H
	Unistat 12
	Uvinul 400
2-hydroxy-4-methoxybenzophenone	Advastab 45
	Cyasorb UV-9
	Uvinul M-40
4-(heptyloxy)-2-hydroxybenzophenone	Unistat 247
2-hydroxy-4-(octyloxy)benzophenone	Carstab 700
2-bydroxy-4-(2-hydroxyethoxy)benzophenone	Eastman Inhibitor HHBP
4-alkoxy-2-hydroxybenzophenone	Advastab 46
	Uvinul 410
2-hydroxy-4-methoxy-5-methylbenzophenone	Unistat 2211
5-benzoyl-4-hydroxy-2-methoxybenzene-sulfonic acid	Cyasorb UV 284
	Uvinul MS-40
2-(2-hydroxy-4-methoxybenzoyl)benzoic acid	Cyasorb UV 207
2,2'-dihydroxy-4-methoxybenzophenone	Advastab 47
	Cyasorb UV-24
4-butoxy-2,2'-dihydroxybenzophenone	Cyasorb UV-287
2,2'-dihydroxy-4-(octyloxy)benzophenone	Cyasorb UV-314
2-(2H-Benzotriazol-2-yl)phenols	
2-(2H-benzotriazol-2-yl)-p-cresol	Tinuvin P
2-$tert$-butyl-6-(5-chloro-2H-benzo-triazol-2-yl)p-cresol	Tinuvin 326
2,4-di-$tert$-butyl-6-(5-chloro-2H-benzo-triazol-2-yl)phenol	Tinuvin 327
2-(2H-benzotriazol-2-yl)4,6-di-$tert$-pentylphenol	Tinuvin 328
Phenyl esters	
phenyl salicylate	Salol
p-(1,1,3,3-tetramethylbutyl)phenyl salicylate	Eastman Inhibitor OPS
resorcinol monobenzoate	Eastman Inhibitor RMB
bis(p-nonylphenyl)terephthalate	Stabilizer BX-721
bis(p-1,1,3,3-tetramethylbutyl)phenyl isophthalate	Santoscreen
Nickel compounds	
bis[2,2'-thiobis-4-(1,1,3,3-tetramethyl-butyl)phenolato] nickel	Ferro AM-101
[2,2'-thiobis[4-(1,1,3,3-tetramethyl-butyl)phenol] ato(2-)] (butylamine) nickel	Cyasorb UV-1084

6.5 Stabilizers against photo-oxidation

Certain common antioxidants used as processing stabilizers, e.g. secondary amines, can absorb light in a region close to the absorption of carbonyl groups, and in the presence of oxygen can themselves give rise to formation of free radicals capable of initiating photo-oxidation of the polymer. Phenolic antioxidants, on the other hand, do not show a strong absorption in this region and can be used as free radical trappers as in the case of thermal oxidation (Chapter 2). Sulphur containing compounds, e.g. di-laurylthiodipropionate, can also be used since they can decompose harmful hydroperoxide groups and exert a synergistic effect in combination with phenolic antioxidants.

Phenolic antioxidants, alone or in combination with thio-compounds, do not provide efficient stabilization against photo-degradation because of the very high rate of the initiation reactions (mechanisms similar to thermal oxidation). Antioxidants alone would be used up readily and extremely high concentrations would be necessary to prevent degradation, (a proposition which is neither technically nor economically attractive). Consequently they are used in conjunction with UV stabilizers to achieve efficient stabilization of the polymer, both in the processing and in service. Hudson and Scott[11] have shown that polymers improperly stabilized against thermal oxidation, which occurs during processing, may degrade much more readily when exposed to UV radiation. They found a direct correlation between the concentration $\geq C = O$ groups with the subsequent UV embrittlement of the polymer.

6.6 Combined effects of temperature, oxygen and UV light

It is quite obvious, from what has been said in Chapter 2 and in the foregoing discussions of this chapter, that any increase in temperature will accelerate further the rate of oxidation of polymers. Owing to their low thermal diffusibility, the surface temperature of plastics components exposed to hot weather conditions can rise much above that of the surrounding atmosphere (a phenomenon which has often been neglected in studies of photo-oxidation).

It is also understood that degradation by photo-oxidation occurs primarily on the surface of the material and invariably fine cracks are formed as a result of natural contractions of the surface layer. These may be formed by the virtue of cross-linking reactions and/or by loss of volatile products.

Under natural weather conditions there is the additional effect of rain water to be considered, which can leach out stabilizers from the outer-surface layer, and in the case of polycondensation polymers they may aggravate degradation by hydrolysis reactions causing a reduction in molecular weight.

6.6.1 Evaluation of the weatherability of plastics

To assess the weathering resistance of plastics both accelerated indoor methods and exposures to natural weather conditions are used. The latter techniques are obviously considered to give more realistic results but are often irreproducible since the weather itself varies so much with the time of the year and location. Factors such as smoke, dust, fog, clouds and the concentration of ozone in the upper atmosphere can all have a considerable effect on the amount of sunlight reaching the surface of the earth.

Accelerated indoor tests are more widely used for both mechanistic studies and formulation work, and artificial light sources which approximate very closely the spectral energy of natural sunlight have been made commercially available in recent years. The Xenon-arc light source filtered through Pyrex glass seems to reproduce very closely the sunlight spectrum in the ultra-violet region, but other sources of UV radiations such as carbon-arcs, fluorescent lamps and mercury-arcs are also frequently used.

Exposures to these sources is often made intermittently in order to simulate day and night conditions and spraying with water droplets is sometimes carried out to reproduce the leaching effect of the additives by rain water.

Appropriate standard test procedures for indoor and accelerated indoor weathering are available.

Hirst and Searle[12] have produced an excellent review of the correlation between accelerated indoor and natural weathering evaluations, and the interested reader is recommended to consult their article.

Irrespective of test method used the process of ageing is invariably monitored by chemical analysis techniques such as infra-red spectroscopy to measure the increase in carbonyl groups, visible light spectroscopy to determine the yellowness index (wavelength range 570–590 nm) and measurements of the destruction of any natural fluorescence or creation of new fluorescent species due to photo-decomposition reactions.

The deterioration of mechanical properties is normally assessed by measurement of fracture resistance by pendulum impact test methods or slow speed flexural strength measurements and ensuring that the surface of the specimens subjected to tensile stresses corresponds to the face of the sample exposed to the weathering source. These tests are generally more meaningful than non-destructive tests, such as modulus measurements, since the chemical changes and the creation of fine cracks (often invisible to the naked eye) may involve only a very fine layer on the surface of the exposed samples.

6.7 Microbiological and hydrolytic degradation

The susceptibility of condensation polymers to hydrolytic degradation has already been mentioned. The extent and mechanism of the degradation depend on the detailed structure of the polymer considered. With cross-linked systems in particular, the process is controlled by the rate of diffusion of water and other accelerating agents, such as bases, acids and micro-organisms.

Hydrolysis of polycondensation polymers is normally accelerated by acid or bases, and the rate of the degradation reactions increases very rapidly with temperature. To protect polymers against hydrolytic decomposition it is essential that any such impurities be removed at the polymer recovery stage of the polymerization process, and that great care is taken when selecting additives in general. When the problem of acid or base catalysed degradation is acute, it may be useful to consider the possibility of using 'buffer' combinations of additives. It is also worth noting that carboxylic acid groups formed as a result of hydrolysis may themselves accelerate further the degradation process[13], producing an autocatalytic effect.

In addition there is some evidence that the carboxylic groups may promote microbiological growth within the system. When this takes place the fungal metabolites may further increase the rate of hydrolytic decomposition.

With hydrolysable nitrogen containing polymers, e.g. polyurethanes, growth of micro-organisms can also take place, which may also increase the rate of decomposition. Other groups which impart biodegradable characteristics to polymer include $-OH$, and $-CHO$.

It has been suggested that 'steric hindrance' in the polymer chains, on the other hand, can contribute considerably to the resistance of polymers to biological attack. It is generally recognized that the problem of biodeterioration of plastics products is more associated with the decomposition of additives rather than of the basic polymer itself. The deterioration of the properties of plastics may therefore, be associated with the loss of additives through microbiological attack[14]. In other words low molecular weight additives, such as lubricants, plasticizers, antioxidants, UV stabilizers, etc., may migrate to the surface of plastics components and encourage the growth of micro-organisms. Removal of such surface fungi by depletion will encourage further diffusion of additives from the bulk and, therefore, devoid the plastics components of some of its vital additives.

Additives which may be used to prevent deterioration of plastics as a result of biodegradation and loss of vital formulations ingredients, are normally called 'fungicides'.

Examples of commercially available fungicides are shown in Table 6.3[15].

Table 6.3

Nature of compound	Applications
Copper 8-quinolinate + toluene sulphonamide condensates	PVC compositions
Phenyl mercury salicylates dissolved in tricresyl phosphate	PVC compositions, acrylics
2,2'-thiobis-(4,6-dichlorophenol)	Polyethylene and polyesters
8-hydroxyquinoline	Polyurethanes

References

1. NEIMAN, N. B., *Ageing and Stabilisation of Polymers*, consultants Bureau (N.Y.), 1965, p. 131.
2. *Modern Plastics*, 9, No. 11, 1963, p. 48.
3. CALVERT, J. G. and PITTS, J. N., *Photochemistry*, Wiley (N.Y.), 1967.
4. HELLER, H. J., *European Polymer Journal*, Supplement 1969, pp. 105–132.
5. *Modern Plastics Encyclopaedia*, McGraw-Hill, 1968, p. 408.
6. BRIGGS, P. J. and McKELLAR, J. F., *J. Appl. Poly. Sci.*, 12, 1968, p. 1825.
7. FINAR, I. L., *Organic Chemistry*, Vol. 1, 3rd Edition, Longmans.
8. LAPPIN, G. R., *Encyclopaedia of Polymer Science and Technology*, Vol. 14, Interscience, 1971, pp. 127–132.
9. MEYER, L. W. A. and GEARHART, W. M., *Ind. Eng. Chem.*, 37, 1945, p. 232.

10. *Encyclopaedia of Polymer Science and Technology*, Vol. 14, Interscience, 1971, p. 140.
11. HUTSON, G. V. and SCOTT, G., *Chemistry and Industry*, 1972, pp. 725–6.
12. HIRT, R. C. and SEARLE, N. Z., *Applied Polymer Symposia*, No. 4, 1967, pp. 61–83.
13. STEINGISER, S., *Rubb. Chem. Tech.*, 37, No. 1, 1964, pp. 38–75.
14. STAUDINGER, J. J. P., *Disposal of Plastics Waste and Litter*, SCI Monograph, NO. 35, 1970, p. 79.
15. RAPRA Information Circular, No. 476, PP. 36–50.

7 OTHER ADDITIVES

7.1 Blowing (or foaming) agents

The function of blowing agents in plastics is to produce cellular materials, i.e. materials containing large proportions of fine cells filled with gas. The embedded cells may be totally enclosed or may be interconnected, for which the terms 'closed cell' and 'open cell' structures are used respectively (Plate 7).

By replacing substantial proportions of solid material by gases, one can utilize to a good measure the properties of both gases and solids in accordance with the laws of mixtures. There may also be some advantages derived from the larger surface areas of gas/solid interface.

The gas properties which are of particular interest in cellular plastics are density, thermal conductivity, dielectric properties, and mechanical and acoustical energy dissipation characteristics.

7.1.1 Effects on density

The law of mixture for density is obeyed exactly and, therefore, the density of cellular materials decreases proportionally to the amount of gas enclosed in the cells:

$$\rho_c = \rho_g \phi_g + \rho_s (1 - \phi_g) \tag{7.1}$$

where ρ_c = density of cellular material
ρ_g = density of the gas phase
ρ_s = density of the solid phase
ϕ_s = volumetric fraction of the gas phase.

Since $\rho_g <<< \rho_s$ the first term can be neglected and the density of the cellular material, therefore, decreases proportionally to ϕ_g.

7.1.2 Effects on thermal properties

The thermal conductivity of gases is substantially lower than that of solids and by arranging the cell structure to be of the 'closed type' and by keeping the size of cells small, convection currents within cellular products can be eliminated. Hence heat conduction through cellular materials may be considered to be equivalent to that of a composite wall, where the gas phase and solid phase are kept at the same temperature.

Hence the overall heat transferred by conduction can be expressed as:

$$Q = \frac{A \, \Delta T}{X} \, [\phi_g K_g + (1 - \phi_g) K_s] \tag{7.2}$$

where
A = interfacial surface area
ΔT = overall temperature gradient
X = respective equivalent thickness for both the gas phase and solid phase
ϕ_g = volumetric fraction of gas
K_g, K_s = respective thermal conductivity of the gas and solid phases.

By making the appropriate substitutions from equation (7.1) and assuming $\rho_g \lll \rho_s$, one obtains

$$Q = \frac{A \, \Delta T}{X \rho_s} \, [\rho_s K_g + (K_s - K_g) \rho_c] \tag{7.3}$$

from which it may be deduced that the heat transfer by conduction increases linearly with the density of the foam.

Radiant heat, on the other hand, is transferred readily through gases, and consequently it is mainly the infra-red absorptivity of the solid phase which determines the resistance to heat flow by irradiation through cellular products. It is to be expected that heat transfer by radiation becomes increasingly more predominant as the density decreases. Owing to radiation reflections from the internal walls of the cells it is difficult to obtain a simple relationship for the heat transfer by radiation through cellular materials but it is likely to follow the curve shown in Fig. 7.1. At very low densities there is a considerable contribution to heat transfer from irradiation transmission through the gas phase, whereas at higher densities the heat conduction contribution through the solid phase becomes the predominant mechanism for the overall heat transfer process[2].

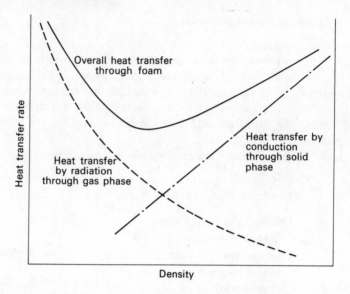

Fig. 7.1 Heat transfer mechanisms through cellular products (idealized behaviour).

7.1.3 Effects on electrical properties

The dielectric properties of materials are related to the polarizability of their internal structure when subjected to an electric field (see Chapter 1 p. 9). Vacuum is a perfect dielectric medium because of the obvious absence of any dipoles. Dielectric losses in gases are also very low because there is little intermolecular friction resistance to the polarization imposed by an applied electric field. Such losses are practically negligible in inert non-polar gases such as nitrogen. With cellular products therefore, neglecting the interfacial effects which may result from adsorption of gaseous material on the inner surfaces of cellular materials, the dielectric constant and dielectric losses would decrease proportionally to the reduction in density.

With cellular polythene, for instance, the Clasius Mosotti[1] relationship is obeyed exactly:

$$\frac{\epsilon - 1}{\epsilon + 2} = k\rho \qquad (7.4)$$

where ϵ = dielectric constant of the material

k = a proportionality constant

ρ = density of the material.

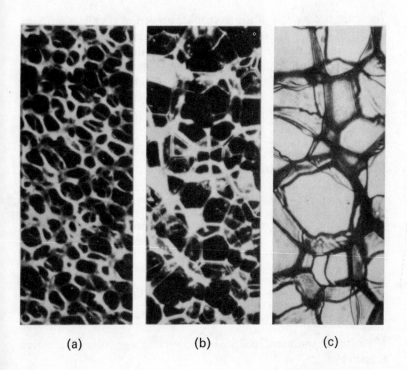

Plate 7 Cellular structure of various plastics foams.
(a) Expanded polystyrene.
(b) Expanded polyethylene.
(c) Flexible polyurethane.
(Reproduced from *Cellular Plastics*, Insulation Products, British Plastics Federation)

Reducing the density of 'low density' polythene from 0.92 to 0.4 g/cm^3 by foaming, results in a corresponding reduction in dielectric constant from 2.29 to 1.4.

7.1.4 Effects on mechanical properties

There are two aspects of the mechanical properties of cellular products which illustrate the beneficial effects of the gas phase embedded in solids, namely rigidity and shock absorption characteristics.

Rigidity considerations can best be made with reference to a cantilever beam, where the maximum deflection is given by:

$$y = \frac{PL^3}{3EI} \qquad (7.5)$$

where P = load applied
     L = length of beam
     E = modulus of the material
     I = cross-sectional moment of inertia

For a beam with a rectangular cross-sectional area, equation (7.5) becomes:

$$y = \frac{4PL^3}{b} \times \frac{1}{Eh^3} \qquad (7.6)$$

where b = width
     h = thickness

If the length, width and load are kept constant the maximum deflection is given by:

$$y = \frac{K}{Eh^3} \qquad (7.7)$$

Assuming that the law of mixtures is obeyed for both density and modulus then a reduction in the density of the material results in a proportional decrease in modulus. So if the weight is to be kept constant, the thickness of the beam would have to be increased by an amount equivalent to the reduction in density; i.e. reducing the density to half of the original value, the thickness of the beam would have to be doubled. Making the appropriate substitutions in the above equation it follows that the maximum deflection of the cellular product would only be one quarter of that of the solid material.

The shock absorption characteristics of cellular products can be described in terms of their cushioning ability, i.e. the ability to stop a falling object with a minimum deceleration. This involves, therefore, considerable absorption of impact energy. If the laws of mixtures for the absorption of energy by the solid and gas phases were obeyed, then replacement of solid material by gas would result in an enormous improvement in shock absorption properties, by virtue of the energy dissipative effects through viscous flow (open cell structures) and compression (closed cell structures) of the gas phase. To avoid damage of cellular products, however, the solid phase would have to deform without breaking or yielding, and consequently the best cushioning effects are always produced with rubbery polymers, where relatively small proportions of gases would be sufficient to produce cellular materials with excellent cushioning properties. If, on the other hand, the breaking strain of the solid phase is low, as in the case of brittle materials such as polystyrene, only small deformations must be allowed and consequently most of the energy will have to be dissipated by the gas phase. Hence large proportions of gases are necessary with rigid foams to produce adequate cushioning properties.

7.1.5 Effects on acoustical properties

There are two aspects of sound insulation which must be considered: insulation of sound generated directly in air and of sound resulting from resonant vibrations of structures.

Cellular materials can be used for both purposes. In the first case the energy from sound waves is dissipated within the cellular material by a viscous damping mechanism of the air, which becomes compressed or moves through intercommunicating cells. Reduction of noise resulting from vibrations of structures, on the other hand, can only be achieved by increasing the stiffness of the structure itself. Since, on a weight basis, cellular products are intrinsically more rigid than the parent solid material, it can be said that a reduction of noise by vibrations can also be achieved by foaming.

7.1.6 Formation of cellular products

The following methods are available to produce cellular materials:

(a) leaching out soluble matter from mixtures or composite materials.

(b) sintering of powders at low pressures in order to leave behind occluded air between sintered particles.
(c) embedding rigid microballoons in a fluid polymeric system and subsequently allowing the surrounding matrix to solidify by either
  (i) formation of chemical cross-links (thermosets),
  (ii) cooling (thermoplastics) or
  (iii) solvent diffusion (PVC pastes)
  This process is used to produce the so-called *syntactic foams*.
(d) By allowing expansion of gases within a fluid matrix in such a manner that gas pockets or cells are formed. Solidification of the matrix is achieved as in (c).
  This is by far the most widely used process to produce plastics cellular products and it is normally known as the 'expansion process'.

The foaming and expansion process

The expansion process for the production of cellular plastics occurs in three successive stages: initiation, growth and stabilization of cells.

(a) *Initiation of nucleation of cells*

Cells are formed when the saturation limits for dissolved gases are reached and if the rate of gas release or expansion within the material is much greater than its rate of diffusion to the surrounding atmosphere. The process of cell nucleation is not fully understood but it is reasonable to assume that cells are formed at those points where solubility is least and where saturation is, therefore, reached first. This deduction can be made from a consideration of Henry's law for the solubility of gases in liquids, i.e. at a given temperature:

$$[C]_T = P[S]_T \tag{7.8}$$

where C = concentration of gas dissolved
  P = external pressure
  S, = solubility coefficient (depending on the nature of the gas and liquid).

Hence at a given temperature and pressure, a reduction in the solubility coefficient will bring about a simultaneous reduction in concentration of gas dissolved, and the excess gas which is expelled will form a bubble.

Therefore, cell formation occurs at points of low gas solubility within the fluid mass, which may consist of finely divided solid inclusions (i.e. pigment particles, fine domains of other low solubility additives, such as lubricants, surfactants, etc.). If cross-linking reactions take place during the generation of gases within the matrix, it is likely that nucleation occurs in domains where the cross-linking density is highest and where solubility is, therefore, least.

With crystalline polymers the heterogeneous nuclei centres for the crystallites could also act as nuclei for the formation of cells in a foaming process.

(b) Growth or expansion of cells

The growth rate of cells in an expansion process is determined by the rate of increase in pressure inside the cells (or by the rate of pressure decay around the cell area) and by the deformability of the cell walls, which can be expressed in terms of surface tension, tensile viscosity or tensile modulus:

$$\left[\frac{\delta r}{\delta t} \right]_T = f \left\{ \psi_{(t)} \left[\frac{\delta(\Delta p)}{\delta t} \right]_T \right\} \tag{7.9}$$

where $\psi_{(t)}$ = deformability function of walls material

   r = cell radius

   Δ_p = net pressure differential between the inside and outside of the cell.

It is likely that nucleation of the cells does not occur simultaneously throughout the matrix and that at any given time during the expansion process, the cell size and the pressure are subjected to normal statistical variations. Owing to surface tension effects there is an excess pressure in smaller bubbles, which causes gas diffusion from one cell to another. Consequently some of the bubbles will grow even larger, some smaller and some may disappear altogether.

(c) Stabilization of cells

It is understood that if the cell growth process was not interrupted at some stage, the surviving cells would grow extremely large and the material forming the walls would reach breaking limits. Eventually all the cells would break into each other and the whole foam structure would collapse.

In the other extreme case, it could happen that all the gas from the cells would slowly diffuse into the atmosphere, the pressure within the cells would gradually decay and the cells would become progressively smaller and disappear if the elastic strain energy in the bulk of the cell walls is the major factor controlling bubble size (e.g. thermoplastics). With very fluid systems (e.g. polyurethanes) and when the cell walls are extremely thin, the factor controlling cell size is the surface tension, in which case a decrease in the excess pressure in the bubbles ($\Delta_p = 2\gamma/r$) would actually cause an expansion of the cells and their subsequent coalescence.

Control of the growth and stabilization of cells is therefore essential in the production of cellular products. This is accomplished by bringing about a sudden solidification or a gradual reduction in deformability of the polymer matrix so that the pressure within the cells will be insufficient to cause further deformation of the walls. With very fluid liquid resins the surface tension of the cell walls is decreased by means of surfactants to encourage growth and stabilization of the cells.

However further reduction in deformability of the matrix, beyond that sufficient for foam stabilization, is necessary to impart mechanical robustness to the cellular product. With thermosetting polymeric systems, both stabilization of the cells and strengthening of the cell walls are achieved by promoting cross-linking reactions in the polymer matrix, whereas with thermoplastics this is achieved by the gradual and controlled cooling of the polymer which increases the viscosity and/or the modulus of cell walls.

A typical nucleation, growth and stabilization mechanism for the production of a thermoset foam is shown in Fig. 7.2. This indicates that, for the formation of a foam having a very fine closed cell structure, it is essential that nucleation coincides with the limits of solubility of the gases evolved in the polymer matrix, and that the molecular weight of the polymer increases very rapidly before the gas evolution rate becomes too low. In this way it is ensured that a positive pressure is maintained in the cells which prevents them from collapsing and at the same time the deformability of the polymer matrix is sufficiently reduced to cause stabilization of the cells. If gas evolution is extremely rapid and the molecular weight of the polymer matrix is too low at the time the saturation limits are reached, the foam will collapse by a wall rupture mechanism. If, on the other hand, gas evolution continues at high rate after nucleation, while the molecular weight of the polymer increases also at a substantial rate, then some wall rupture will take place forming an irregualr open cell structure, but collapsing will be prevented by the fairly low deformability achieved by the polymer at this stage.

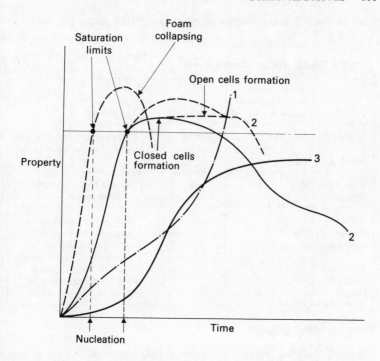

Fig. 7.2 Idealized expansion process: curve 1 = molecular weight of polymer matrix; curve 2 = gas evolution; curve 3 = foam volume.

7.1.7 Additives used in expansion processes

Foaming agents used for the production of cellular plastics are normally divided into physical and chemical types according to whether the generation of gas is by a physical process (i.e. evaporation or sublimation) or by a chemical process (i.e. breakdown of the chemical structure or by other chemical reactions).

Physical blowing agents

These are normally low boiling liquids or gases which are soluble in the polymer matrix and exert their blowing action when brought to boiling conditions by increasing the temperature and/or reducing the pressure of the system.

The chemical nature and the deformability characteristics of polymer systems determine the type of physical blowing agent that may be used. Some examples of physical blowing agents are shown in Table 7.1.

Table 7.1 Typical physical blowing agents[3].

	Boiling range at atmospheric pressure (°C)	Comments	Typical uses
Pentane	30–38	Non-toxic	Polystyrene foams, (mainly by expandable beads route)
Heptane	65–70	„ „	„ „
Methylene chloride	40	Toxic, non-flammable	Flexible PVC and polyurethane foams
Trichloro-fluoro-methane	24	Non-toxic non-flammable	Most widely used foaming agents. Major applications include polyurethane (flexible and rigid) foams and PVC flexible foams. Also used as auxiliary to chemical blowing agents.
Trichloro-trifluoro-ethane	47	„ „	

Chemical blowing agents

Chemical blowing action in thermosetting materials often occurs through formation of volatile by-products of chain extension and cross-linking reactions, as in the case of polyurethanes, phenolics and aminoplastics. With thermoplastics, on the other hand, chemical blowing is achieved by means of compatible or finely dispersed chemical substances decomposing at a required rate over a fairly narrow temperature range, which is above the compounding temperature but within the processing temperature range of the polymer.

The most important chemical blowing agents used commercially are compounds which release nitrogen as the major component of the gas phase (Table 7.1). Furthermore the decomposition reactions can be catalysed or inhibited in some cases, so that the blowing rate can be adjusted in any given formulation to suit the particular processing conditions used (pressure, temperature and time). Typical chemical blowing agents found commercially are shown in Table 7.2.

7.1.8 Evaluation of blowing agents

Irrespective of whether cellular materials are produced by chemical or physical blowing it is necessary to have a knowledge of the solubility

Table 7.2 Typical chemical blowing agents for cellular plastics[4]

Name	Structure	Decomposition range ($^\circ$C) (maximum rate)	Comments
Azodi-carbonamide	$NH_2-CO-N=N-CO-NH_2$	160—200	Non-toxic. Suitable for polymers processed at fairly high temperatures, can be activated.
Azobisbutyronitrile	$\begin{array}{cc} CH_3 & CH_3 \\ NC-C-N=N-C-CN \\ CH_3 & CH_3 \end{array}$	90—115	Gives toxic residues Can be activated or inhibited.
Benzene-sulphonyl-hydrazine	$\langle\bigcirc\rangle-SO_2-NH-NH_2$	95—100	Limited applications
P-toluene Sulphonyl semi-carbazide	CH_3 ... $SO_2-NH-NH-CONH_2$	210—270	Suitable for thermoplastics processed at high temperatures, e.g. nylons, polypropylene, etc.

characteristics of the gases, which cause the expansion, under actual processing conditions. In the case of chemical blowing agents compatibility of the agent with the polymer prior to its decomposition is essential in order to achieve a uniform nucleation during the expansion process. Consequently for the purpose of developing potential chemical blowing agents, one would be concerned with measurements of the rates of gas evolution by TGA and DTA, assessment of the compatibility in the polymers considered and determinations of the solubility of the mixtures of gas evolved at various temperatures and pressures using appropriately constructed leak-proof apparatus. Furthermore one would have to consider the toxicity hazard of any volatile by-products formed and the possible interactions of these with other ingredients such as pigments.

For the development of cellular products, on the other hand, it is somewhat more difficult to specify general evaluation procedures owing to the multitude of polymeric compositions and processing variables which would have to be taken into account. A pre-selection of a suitable blowing agent can be made, however, on the basis of its rate of decomposition and gas evolution over the temperature range applicable to the process considered. From these it would be possible to estimate the approximate concentration of blowing agent required to produce a foam of a predetermined density, using the law of mixtures on p. 146. For the case of injection moulded or extruded thermoplastics foams it would be desirable to have a knowledge of the solubility characteristics of the gases evolved in order to estimate the pressure requirements to keep these dissolved in the melt. Insufficient pressure, in fact, will cause the formation of premature gas bubbles and result in uneven cell growth[5].

One would ultimately use these estimates as a basis for the selection of levels of additives and range of processing conditions to be used in the evaluation programme.

7.2 Flame retardant agents

7.2.1 Burning mechanism of plastics

All polymers used in plastics products to date are thermally unstable, i.e. on heating they undergo chemical breakdown with formation of volatiles.

If plastics materials are therefore exposed to heat in air, degradation reactions will take place and the volatiles formed may leave behind a porous residue. This will ease the penetration of oxygen from surrounding air and cause further oxidation reactions in the solid substrate. The residue often consists of a carbonaceous char, which increases the amount of heat absorbed from surrounding radiations, and promotes further pyrolysis of the material beneath. There will be therefore a cumulative rise in temperature and eventually the vapours will ignite to form a flame. Ignition may be induced by an external flame (flash-ignition) or may occur spontaneously (self-ignition). The heat generated by combustion may sustain the ignition by continually providing the necessary thermal energy for the pyrolysis of the substrate material. In this case the material is said to be *flammable*. Such heat of combustion may not be sufficient, on the other hand, to provide the necessary thermal energy to cause the material to pyrolyse and produce volatiles at a sufficient rate to sustain ignition, and consequently the flame will eventually extinguish. In this case the material is said to be *self-extinguishing*.

The temperature at which volatiles produced from pyrolysis will ignite and the subsequent combustion rate depend on their chemical constitution and on the relative proportions of combustible volatiles and oxygen present. Ignition will not take place nor will the flame be sustained if the concentration of combustible volatiles is below or above their flammability limits

7.2.2 Chemistry of combustion of volatiles

Volatiles from thermal degradation of polymers invariably contain C–C and/or C–H groupings which constitute sites for free radical formation and oxygen attack.

$$CH_2 \xrightarrow{heat} CH^* \xrightarrow{O_2} CHOO^* \longrightarrow CHO + HO^*$$

$$\begin{array}{l} + \\ H^* \xrightarrow{O_2} \begin{array}{l} HOO^* \\ HO^* \\ O^* \end{array} \quad \begin{array}{l} (\text{wall effect} \\ + H_2 \longrightarrow H_2O \text{ etc.} \\ HO^* \end{array} \end{array}$$

$$H_2 \xrightarrow{O_2} H_2O$$

It appears that the ignition susceptibility and the rate of flame propagation are related to the ease and the rate of formation of OH* radical[5]. The reactions leading to the formation of CO_2 and H_2O are highly exothermic and the heat developed provides an autoaccelerative effect on the oxidative ractions and on the final rate of combustion. The presence of solid particles in the gas phase, however, can reduce to some extent the excitation energy for the oxidative process by the so-called 'wall effect', so that larger amounts of the less active *OOH radicals are produced. There may also be some contribution from charred residues which may burn by glowing and generate more heat.

The flammability characteristics of plastics depend, therefore, on (i) their basic chemical structure, (ii) the intrinsic flammability of volatiles and (iii) the ratio of combustible matter in the gas phase.

The relation of the products of pyrolysis and combustion to the ignition susceptibility and the burning rate of the most common plastics materials is shown in Table 7.3. Non-combustible products of pyrolysis such as halogen halides, amines, CO_2, fluorocarbons, etc. have the effect of raising the ignition temperature, increasing the oxygen demand for sustained ignition, and may cause self extinction of the burning products.

Table 7.3 Relationship between structure, pyrolysis products and intrinsic flame retardancy of common plastics[1].

Material	Pyrolysis products	Combustion products	Flash[a] ignition temp. (°C)	Limiting[b] oxygen for sustained ignition (%)	Burning[c] rate cm/min
Polyolefins	Olefins, paraffins, alcyclic hydrocarbons.	CO, CO_2	343	17·4	1·75–2·5
Polystyrene	Styrene monomer, dimers, trimers.	CO, CO_2	360	18·3	2·5–3·7
Acetals	Formaldehyde.	CO, CO_2		16·2	≅ 2·5
PTFE	Fluoro-hydro-carbon monomers.	— —	—		NB
Acrylics	Acrylate monomers.	CO, CO_2	338	17·3	1·2–5·0
PVC	HCl, aromatic hydrocarbons.	HCl, CO, CO_2	454	47·0	SE
Tere-phthalates	Olefins, benzoic acid.	CO, CO_2			
Cellulose acetate	CO, CO_2, acetic acid.	CO, CO_2, acetic acid	327	25·0	1·2–5·0 SE
Poly-carbonates	CO_2, phenol	CO, CO_2	482		SE
Nylon 66	Amines, CO, CO_2.	CO, CO_2 NH_3, amines	424	28·7	SE
Phenolics	Phenol, formaldehyde.	CO, CO_2 HCOOH	482	—	SE
Melamines	NH_3, amines.		602	—	SE
Polyesters	Styrene, benzoic acid etc.	CO, CO	485	—	≅3·7

Notes:
(a) ASTM D1929-62T
(b) Candle-Type Test, (Modern Plastics Nov. 1966, p. 141)
(c) ASTM D635
SE = Self-extinguishing
NB = Non-burning

7.2.3 Approach to fire retardancy of plastics

By their very nature, polymeric materials cannot be made 'fire proof', hence the only open possibility is to reduce their susceptibility to incipient and sustained ignition. Agents capable of reducing the ease of ignition and the rate of flame propagation are called *fire-retardants*, and the level of reduction in flammability is defined according to test specifications.

Four possibilities are available for promoting fire retardancy in plastics[6].

(1) Coating the exposed area to reduce oxygen permeation, hence decreasing the rate of oxidative reactions.

 If the protective coating is intumescent it would provide an effective thermal insulation layer and reduce the rate of formation of volatiles by pyrolysis of the substrate.

(2) Forming large amounts of incombustible gases which would dilute the oxygen supply and reduce the rate of combustion. This in turn would decrease the temperature of the material, which may fall below the ignition temperature and cause self extinction. Examples of such gases are ammonia, nitrogen, sulphur dioxide and hydrogen halides.

(3) Promoting endothermic reactions in the exposed regions in order to reduce the temperature below that which would sustain ignition. One effective way of achieving this is to use substances which sublime at temperatures around ignition temperature of the material.

(4) Inhibiting the free-radical oxidation process so that the rate of formation of very active OH* radicals is reduced. This can be achieved by trapping radicals or deactivating them through chemical reactions and/or wall effects.

An important characteristic of flame retardant additives is their efficiency which is independent of their state of dispersion or solubilization in the plastics matrix. Most of the flame retardation reactions occur in fact in the gas phase and, therefore, it is the rate of diffusion of the inhibiting species and/or their subsequent rate of reactions with the free radicals which mostly determine their effectiveness.

Protective and intumescent coatings

These additives are used on the outer surface of plastics components, normally applied in the form of paints.

The additive must be able to produce incombustible gases at a very high rate so that they will foam the outer surface and expel oxygen.

The foam also provides a thermal insulation barrier to the substrate. The coating material is normally a polyalcohol[7] (sugars, starch, sorbitol, pentaerythritol, etc.) which is mixed with a blowing agent and a dehydrating agent. The blowing agents normally used are melamines, urea, guanidines, etc., which give out incombustible volatiles by decomposition.

The dehydrating agent is a Lewis acid which promotes charring of the polyol by an internal cyclization mechanism:

(a) $R(OH)_2 + Acid \longrightarrow RO + Acid \cdot H_2O$

simple dehydration

or

(b) $R(OH)_2 + Acid \longrightarrow [R(OH)_2 \cdot Acid] \longrightarrow RO + Acid + H_2O$ gas

The acid most widely used is phosphoric acid introduced in the form of a low water-solubility phosphate, e.g. esters and ammonium or amine phosphates. Phosphoric acid is formed by decomposition of the phosphate when the coating is exposed to high temperatures. A coating material may, however, be formed by the melting and fluxing of incombustible additives, e.g. borates, which can form a barrier to oxygen penetration.

Flame retardants functioning by the formation of incombustible products

There are only a few cases of flame retardants which function by this mechanism alone. These include ammonium sulphate and ammonium sulphamate which decompose into ammonia, water and SO_3/SO_2 on heating. They are often used in papers and other cellulosic products, but rarely in plastics as they are not sufficiently stable at processing temperatures and would leach out in service.

Flame retardants functioning as free-radical trappers

There are two classes of flame retardants which exert their function as free-radical trappers, namely organic bromides and chlorides.

At high temperatures these decompose to yield hydrogen halides which react preferentially with the very active *OH radicals[5].

$$R-X \xrightarrow{\text{heat}} R^* + X^*$$

$$+$$

$$R'H \longrightarrow HX + R'^*$$

$$+ {}^*OH$$

$$\longrightarrow H_2O + X^*$$

The efficiency of these compounds is therefore related to the ease of dissociation of their $C - X$ bonds. The rate at which the halogen radicals are formed must also be such that a continuous source of radical trappers is available for as long as the temperature of the exposed region is above the ignition temperature of the volatiles.

Hence, although bromine compounds have a lower bond dissociation energy than the correspondent chlorides (Table 7.4), they are not necessarily more efficient flame retardants.

Table 7.4 Bond dissociation energies of organic halogen compounds[8] (kJ/mol).

R	—Cl	—Br
H—	420	370
CH_3—	340	285
$(CH_3)_2 CH$—	340	280
$(CH_3)_3 C$—	320	275
C_6H_5—	360	298
$C_6H_5 CO$—	310	240

Typical halogenated organic compounds used as flame retardants are shown in Table 7.5.

Table 7.5 Chlorohydrocarbons used as fire retardants in plastics.

Compound	Suggested uses
Chlorinate paraffins ($C_{10}-C_{30}$)	General usage
Hexachlorocyclopentadiene derivatives	,, ,,
Chlorinated aryl diamines	Epoxy resins
Chlorinated alkyl aryl ethers	Polyesters
Tetrabromododecene	,,
Hexabromocyclopentadiene	,,
Pentabromtoluene	Polyurethanes
Bromophthalimide	Nylons

Multimechanistic and synergistic flame retardant systems

Most flame retardant agents exert their action by more than one of the mechanisms previously described, and mixed systems are often used to obtain synergistic effects. The most common fire retardant agents used commercially are:

(a) Compounds containing halogen and phosphorus. The advantage of these compounds is that they can exert a radical trapping mechanism through reactions with halogen radicals and, at the same time, phosphoric acid residues formed can promote formation of char by dehydration of alcohols. Smoke will also be formed as a result, which can contribute to deactivation of radicals by the wall effects. It is also possible that phosphorous pentoxide may be formed, which would dilute the combustible volatiles and expel oxygen from the surface of the material.

(b) Halogen compounds/antimony oxides mixtures. Antimony oxide offers virtually no flame retardation by itself since it melts at temperatures above the ignition temperature of most plastics and, therefore, cannot afford fire protection even by a coating mechanism. In mixtures with halogen compounds however it can form antimony chlorides and oxychlorides, which are gaseous at the ignition temperature and will dilute the combustible gases.

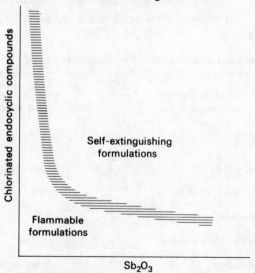

Fig. 7.3 Synergistic behaviour of binary mixtures of Sb_2O_3/hexachlocyclopentadiene compounds (idealized curves).

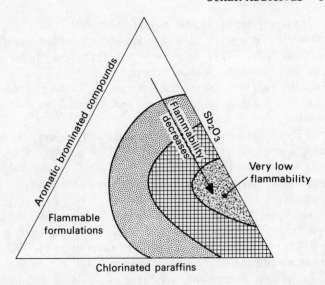

Fig. 7.4 Synergistic behaviour of ternary mixtures of Sb_2O_3/chlorinated paraffins/brominated compounds[10] (idealized curves).

In addition, reactions occurring in the solid phase can promote char formation in a manner similar to phosphoric acid (Fig. 7.3 and Fig. 7.4).

(c) Boron compounds and mixtures. Boric acid and derivatives have fluxing characteristics i.e. they melt and form a protective film. Boric and metaboric acids melt and flux in the temperature range 170 °C to 250 °C, which is below the ignition temperature of most plastics and, therefore, can be usefully used as flame retardants. Zinc and calcium borate has been used in paints but no details are available as to the mechanism of their flame retardant action. They suffer from the drawback of being highly hygroscopic.

High molecular weight organo-borates compounds have been introduced recently and have the advantage of greater combatability in polymeric systems and lower extractability than inorganic compounds.

7.2.4 Evaluation of fire retardant additives

Whereas the chemical techniques to study the mechanisms of flame retardants are reasonably well established, small scale test methods which will provide meaningful data to predict their performance under conditions of actual use are still very much disputed.

Basically the tests are devised to provide information regarding (a) conditions under which the material will ignite and (b) the rate at which the same will be expected to burn.

For the purpose of selecting additives and establishing suitable levels of addition in a given formulation the recommended test methods are as follows:

(a) Oxygen index determinations and ignition temperature measurements, to assess the ignition susceptibility of the material.

(b) Rating the material as 'self-extinguishing' or 'flammable' and measuring the 'surface spread of flame', to assess the resistance to fire propagation.

The oxygen index, n, is defined as the minimum concentration of oxygen in a nitrogen/air mixture which supports ignition of a candle-like specimen. The greater the value of n the higher the resistance of the material to support ignition.

The ignition temperature, on the other hand, is defined as the temperature at which the material undergoes either spontaneous ignition (self-ignition temperature) or flame induced ignition (flash-ignition temperature) when a sample is subjected to a standard rate of temperature rise and air supply.

A material is deemed to be 'self extinguishing' when a horizontal strip inclined at $45°$ and exposed for 30 seconds to a $2·5$ cm flame fails to burn beyond 1 cm distance from the ignited edge. The rate of burning, on the other hand, is determined by igniting a plaque with its edge placed at an angle to a source of radiant heat which gives a temperature gradient along the specimen of approximately $10 °C/cm$.

All these tests are covered by standard test methods where the interested reader can find all the details of the appropriate procedures.

It is understood, however, that ultimately the acceptance of a flame retardant material for a given application will be determined by its satisfactory performance in special test specifications devised by appropriate authorities.

References

1. FREEMAN, W. R., *Electrical Manufacture*, August, 1962.
2. FERRIGNO, T. H., *Rigid Plastics Foams*, 2nd Edition, Reinhold, 1967, p. 257.
3. *Modern Plastics Encyclopaedia*, McGraw-Hill, 1968, p. 369.
4. *Ibid*, p. 370.

5. COLLINGTON, K. T., *Plastics & Polymers*, February, 1973, p. 24.
6. SCHMIDT, W. G., *Trans. Plast. Inst.,* 33, 1965, p. 248.
7. BELL, K. M., The Plastics Institute Conference on Flame Resistance with Polymers, London, 1966.
8. LYONS, J. W., *The Chemistry and Uses of Fire Retardants*, Interscience, 1971.
9. JOLLES, Z. E., The Plastics Institute Conference on Flame Resistance with Polymers, London, 1966.
10. Berk Ltd., Flammex 5BT, Publication P701/2.

INDEX